AF597917

Electrophoresis - Recent Advances, New Perspectives and Applications

Edited by Yee-Shan Ku

Published in London, United Kingdom

Electrophoresis - Recent Advances, New Perspectives and Applications
http://dx.doi.org/10.5772/intechopen.108843
Edited by Yee-Shan Ku

Contributors
Hui Li, Jiaquan Xu, Ahmed Khiredine Metref, Aida Meto, Agron Meto, Estefanía García-Luque, Ana del Pino-Pérez, Enrique Viguera, Cheuk-Hin Law, Long-Yiu Chan, Tsz-Yan Chan, Yee-Shan Ku, Hon-Ming Lam

First published in London, United Kingdom, 2024 by IntechOpen
IntechOpen is the global imprint of INTECHOPEN LIMITED, registered in England and Wales, registration number: 11086078, 5 Princes Gate Court, London, SW7 2QJ, United Kingdom

British Library Cataloguing-in-Publication Data
A catalogue record for this book is available from the British Library

Additional hard and PDF copies can be obtained from orders@intechopen.com

Electrophoresis - Recent Advances, New Perspectives and Applications
Edited by Yee-Shan Ku
p. cm.
Print ISBN 978-1-83769-431-0
Online ISBN 978-1-83769-430-3
eBook (PDF) ISBN 978-1-83769-432-7

For EU product safety concerns:
IN TECH d.o.o., Prolaz Marije Krucifikse Kozulić 3, 51000 Rijeka, Croatia,
info@intechopen.com or visit our website at intechopen.com.

Meet the editor

Dr. Yee-Shan Ku received her Ph.D. in Molecular Biotechnology from The Chinese University of Hong Kong. She is currently a post-doctoral researcher in the School of Life Sciences and Centre for Soybean Research of the State Key Laboratory of Agrobiotechnology, The Chinese University of Hong Kong. Dr. Ku started her research with functional studies of plant genes. Her research focuses on plant stress adaptability and signaling, plant metabolite, and plant–microbe interaction. She is interested in comprehending life sciences at the molecular level. Dr. Ku is also an active reviewer and editor for international peer-reviewed journals.

Contents

Preface

Electrophoresis is a classic molecular biology technique for analyzing biomolecules including DNA, RNA, protein, and their complexes. The straightforward yet effective idea of electrophoresis is to physically separate biomolecules according to their sizes and charges. Combined with other technologies such as chromatography, mass spectrometry, and nucleic acid sequencing, electrophoresis has been applied to achieve various purposes such as sample purification, molecule identification, and species identification.

This book showcases the applications of electrophoresis in a broad spectrum of research fields. The highlighted research fields include single-cell technology, veterinary diagnosis, dental research, biodiversity study, and soil research. Single-cell technology features a high resolution of an individual cell. Due to the small amount of sample input, delicate products are required to prepare the samples for downstream analyses. Examples of electrophoresis-aided procedures include molecule separation and ionization. In veterinary diagnosis, electrophoresis is an important tool for analyzing serum protein fractions. The serum protein profile is then used for animal health assessment. In the field of dentistry, electrophoresis has been widely employed to evaluate dental cement and composite materials. Moreover, electrophoresis is applied to detect dental disease biomarkers.

Electrophoresis is also applied in biodiversity study and soil research. Coupled with another classic molecular biology technique, polymerase chain reaction (PCR) and DNA sequencing, electrophoresis is involved in the generation of molecular barcodes for diverse species. For soil research, coupled with other techniques such as chromatography, mass spectrometry, PCR, and DNA sequencing, electrophoresis is useful for analyzing compounds, microbes, plants, and animals in soil.

Electrophoresis is a versatile technique suited for various purposes. In addition to summarizing the current applications of electrophoresis in different research fields, we hope this book will inspire readers to unravel the potential and appreciate the beauty of this classic technique.

Yee-Shan Ku
School of Life Sciences,
Centre for Soybean Research,
State Key Laboratory of Agrobiotechnology,
The Chinese University of Hong Kong,
Hong Kong SAR, China

Chapter 1

Application of Electrophoresis in Single-Cell Analysis by Mass Spectrometry

Hui Li and Jiaquan Xu

Abstract

The cell serves as the fundamental building block of life, yet it operates as an extremely sophisticated chemical system. Single-cell analysis holds the potential to provide novel insights into cellular heterogeneity and their corresponding subpopulations at the genomic, transcriptomic, proteomic, and metabolomic levels. Mass spectrometry (MS) is a label-free technique that enables the multiplexed analysis of proteins, peptides, lipids, and metabolites in individual cells. By now, the application of electrophoresis in single-cell analysis by MS has become widespread. In this chapter, we will summarize the recent application advancements of electrophoresis in single-cell analysis by MS, with a particular focus on sampling, separation, and ionization. Additionally, we will discuss potential future research directions for utilizing electrophoresis in single-cell analysis by MS.

Keywords: single-cell analysis, capillary electrophoresis mass spectrometry, ambient ionization, single-cell sampling, separation

1. Introduction

Cell is the basic unit of the organism's structure and function [1], which has a close relationship with the occurrence, development, and treatment of diseases. Individual cells exhibit significant diversity due to a variety of factors, including genetic variations, fluctuations in biochemical processes, and differences in their microenvironment. Despite the homology between two cells, there can be significant variations in their intracellular material composition and content [2]. The conventional assessment of cell population can solely provide an average outcome regarding homeostasis and fails to depict the variances among individual cells. Consequently, single-cell investigation not only enhances our comprehension of cellular nature and life but also furnishes a more potent mechanism for disease diagnosis, categorization, therapy, and prognosis. However, the minute dimensions of the cells pose a challenge in terms of isolating individual cells and preparing samples for analysis. Due to the minuscule size, substances present within single cells are extremely scarce and exhibit significant concentration disparities spanning nine orders of magnitude [3], thereby exacerbating detection difficulties.

Among the single-cell analysis methods, fluorescence detection methods are highly sensitive and can detect single molecules with dynamic tracking ability [4–7], but their detection flux is limited. Electrochemical detection has high temporal-spatial resolution [8–11], but it cannot detect multiple molecules simultaneously due to the limitations imposed by the electrochemical window and the requirement for electrochemically active molecules. Mass spectrometry (MS) is a conventional method for molecule analysis with the advantages of high sensitivity, high throughput, and the ability to identify molecular structures [12, 13]. By leveraging an existing database, this method can qualitatively and quantitatively analyze tens of thousands of molecules at the same time while providing abundant information on their structure characters. However, the application of direct MS in single-cell detection faces challenges in terms of sample complexity and single-cell sensitivity due to the diverse range and limited content of chemical components within cells. Therefore, it is imperative to employ appropriate separation technology to improve the method's detection sensitivity and qualitative/quantitative accuracy.

Typically, in various omic studies, liquid chromatography (LC) and its derivatives, such as high performance liquid chromatography (HPLC), ultra performance liquid chromatography (UPLC) and nano liquid chromatography (nanoLC), are common separation techniques [14, 15]. However, LC may not be suitable for samples with limited volume due to the requirement of a larger injection volume (typically several microliters) [16]. On the contrary, electrophoretic separations like capillary electrophoresis (CE) possess distinct advantages compared to chromatographic methods. Since the late 1980s, CE has emerged as an exciting and promising microelectric separation technique due to its benefits such as minimal sample consumption, high resolution, rapid separation time, and cost-effectiveness. Various modes of CE including capillary zone electrophoresis (CZE), capillary isotachophoresis (CITP), capillary isoelectric focusing (CIEF), micellar electrokinetic capillary chromatography (MEKC), and capillary electrochromatography (CEC) have proven beneficial in pharmacology, food analysis, biomarker research, and biomolecular analysis [17, 18].

In this chapter, we will provide a comprehensive overview of the application of electrophoresis in single-cell MS, encompassing three main aspects. The first part starts with an introduction to the application of electrophoresis technology in single-cell sampling for MS analysis. The second part further describes how to utilize electrophoresis technology to separate the components within single cells, which are subsequently analyzed using MS. The final part provides a detailed explanation of the application of electrophoresis in the ionization source, enabling simultaneous separation and ionization of components within single cells.

2. Application of electrophoresis in sampling single cells for MS analysis

Sampling analytes from a single cell in MS analysis is a highly challenging task due to the small sizes of cells (usually several tens of micrometers in diameter for mammalian cells), limited cellular content volume, complex intracellular matrix, and rapid turnover. As a result, there have been numerous advancements aimed at enhancing sampling and extraction efficiency before conducting MS analysis.

2.1 Sucking contents from single cells for MS analysis

Sucking contents from individual cells for MS analysis is a commonly employed technique, typically involving the use of a nanospray/capillary tip to extract cell contents through either negative pressure or cellular pressure. For instance, Mizuno et al. devised a video-MS method for real-time single-cell analysis. In their investigation, they utilized a micromanipulator-mounted nanospray tip (with a diameter of 1–2 μm and coated with gold) to access the cytoplasm or organelles of live mammalian cells under microscopic observation. By connecting a tube-connected piston syringe, several hundreds of femtoliters of cellular contents were aspirated into the nanospray tip during sampling procedures [19]. To achieve direct analysis of target organelles in single cells, a method by combination of fluorescence imaging and live single-cell MS was developed Esaki et al. [20]. Initially, a mitochondria-specific fluorescent marker was utilized to visualize the location and condition of mitochondria within living cells. Subsequently, the stained mitochondria were selectively captured with a nanospray tip under fluorescence microscopy, enabling predominantly metabolite detection by live single-cell MS. Zhang et al. [21] utilized capillary microsampling in combination with electrospray ionization (ESI) MS and ion mobility separation to conduct metabolic analysis of various types of epidermal cells in *Arabidopsis thaliana*. Their research involved the use of a nanospray tip (with a diameter of 1 μm, uncoated with metal) to extract cytoplasm from individual plant cells through capillary action and turgor pressure.

The precise control and measurement of the volume sampled from single cells is essential for accurate quantitation of the analytes at the single-cell level. Thus, Yin et al. developed a technique for sampling single cells using electroosmotic extraction with meticulous control [22]. The extraction nanopipette, with a tip diameter of less than 1 μm, was prepared by using a capillary puller, followed by the insertion of a Pt electrode. A counter electrode Pt was connected to the sample. In a typical sampling process (**Figure 1**), the nanopipette was filled with 5 μl of hydrophobic electrolyte solution, and during the piercing procedure, a voltage of +2 V (with electrode 1 as the positive pole) was applied between the two electrodes to effectively prevent solution

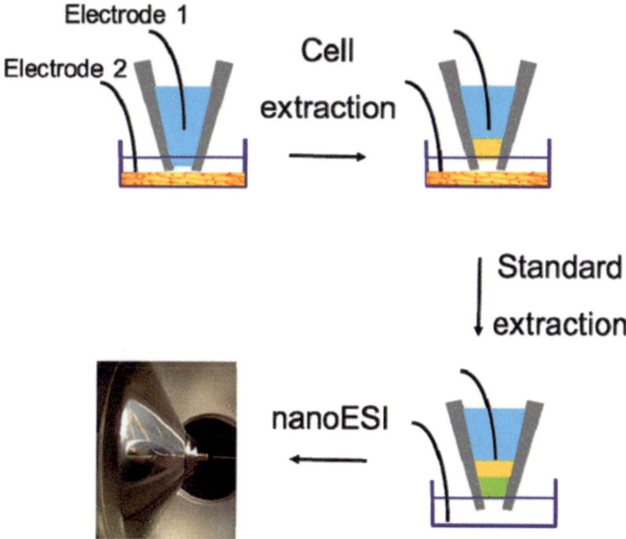

Figure 1.
Schematic diagram of electroosmotic suction of single cells for MS analysis. Electrode 1 was inserted into the nanopipette, while Electrode 2 was connected to the sample. The initiation of sucking contents from a single cell (yellow) involved applying a negative voltage between Electrode 1 and Electrode 2. Subsequently, a solution containing an internal standard (green) with a known volume was introduced into the nanopipette using electroosmotic extraction. Finally, the extracted analytes were analyzed by nESI-MS [22].

aspiration into the tip. The nanopipette was then guided toward the cell at a speed of 400 nm/s using micromanipulation and microscopy while monitoring ion current between the electrodes. The moment the cell membrane was approached, a discernible modification in ion current became evident. After inserting the nanopipette into the cell, a voltage of −2 V (with electrode 1 as the negative pole) was applied between the two electrodes. This allowed for precise control over the suction volume of cytoplasm, facilitating subsequent analysis through nanoelectrospray ionization (nESI) MS. By incorporating an isotope internal standard, this technique facilitated quantitative analysis of metabolites from individual cells.

2.2 Extraction of analytes from single cells for MS analysis

The extraction strategy of analytes from single cells for MS analysis can be categorized into two methods. One method utilizes the solid-liquid microextraction (SLME) technique, while the other relies on liquid-liquid microextraction (LLME). The single-cell MS analysis techniques based on SLME involve probe electrospray ionization (PESI)-MS [23], direct sampling probe (DSP)-MS [24], and surface coated probe (SCP)-nESI-MS [25], and all of these procedures involve the utilization of a metal needle that has undergone surface treatment/modification/coating, which is then employed for the purpose of extracting and enriching analytes from an individual cell. The advantages of SLME are high spatial resolution and selectivity, while the disadvantage of SLME is low throughput. Nanomanipulation-coulped nanospray MS [26] and single-probe MS [27] are typical LLME-based single-cell MS techniques, which involve the utilization of solvent to extract and isolate analytes from individual cells. The strength of SLME lies in its high-throughput capability, while its limitation lies in the relatively lower level of selectivity.

In cellular environments, the majority of biomolecules possess either positive or negative charges, which are determined by their biophysical characteristics and surrounding conditions. Consequently, when exposed to an external electric field, these charged biomolecules have the ability to migrate. Positively charged biomolecules tend to move toward the cathode, while negatively charged metabolites tend to move toward the anode, resulting in a polarity-specific separation process. As a result of this phenomenon, Song et al. [28] introduced probe electrophoresis as a sampling technique for single cells (**Figure 2**). To prepare the probe, a stainless steel needle coated with gold and featuring a 100 nm tip was inserted into a capillary with a 2 μm tip that had been pulled by laser. The distance between the tip of the needle and capillary was approximately 5 μm. Before the sampling procedure, the single cells were already deposited onto an indium-tin oxide (ITO) substrate that had been positioned correctly under the microscope. The probe containing 9 pl of water was then carefully inserted into a single cell using micromanipulation techniques under the microscope. After remaining in a state of quiescence for several seconds, during which any potential perturbations were eliminated, low DC voltage (±2 V) between the probe and conductive substrate was applied to generate strong electric fields (around 10^6 V/m) inside the cell due to their close proximity (about 5 μm).

An electric field gradient with a magnitude of 10^{19} V^2m^{-3} was created near the tip of the probe. This resulted in the movement of biomolecules with opposite charges in different directions: either upward toward the probe or downward toward the conductive substrate. To prevent electroosmotic flow, the back end of the probe was subjected to air pressure generated by a microinjector during electrophoretic sampling. Following extraction, the probe was taken out from the cell and directly

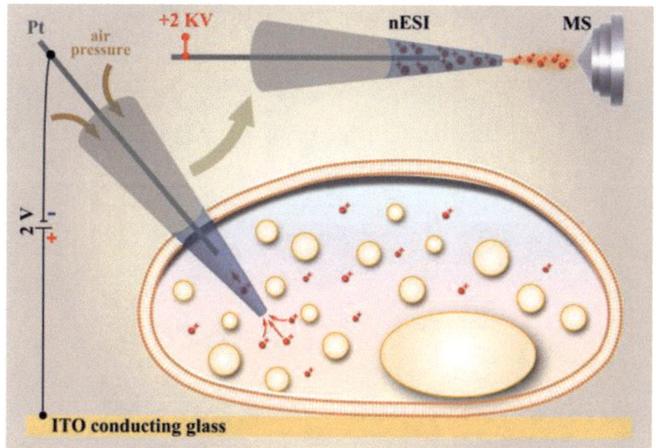

Figure 2.
The schematic diagram of single-cell analysis by probe electrophoresis MS. The charged metabolites in a single cell were first extracted via electrophoresis by applying a voltage between Pt electrode and ITO conduction glass, followed by nESI-MS analysis. [28]

utilized as a nESI emitter. The application of a high voltage (±2 kV) to the probe facilitated the generation of analyte ions for MS analysis. The findings indicated that the extraction efficiency was influenced by factors such as the magnitude and duration of extraction voltage, concentration of analytes, and dielectric constant of the electrolyte. Under optimal conditions, the method can achieve high throughput (≥200 peaks) and enhanced sensitivity (≥10-times signal enhancement for $[Choline+H]^+$, $[Glutamine+H]^+$, $[Arginine+H]^+$, etc.) compared to direct nESI-MS.

3. Application of electrophoresis in separation of the analytes from single cells for MS analysis

The enhanced resolving abilities of CE have been effectively employed in intriguing biological studies conducted at the individual cellular level. These advancements in CE have significantly broadened our understanding of metabolites, proteins, and other relevant components. Currently, numerous literature reviews have comprehensively documented the utilization of CE-MS for analyzing single cells [29, 30].

3.1 Metabolites

Cellular metabolic analysis plays a crucial role in the examination of cells, offering insights into the physiological state of organisms and providing a fundamental understanding of their overall health. By utilizing the CE-MS technique, researchers have successfully established cell metabolism analysis as an effective tool for addressing various biomedical and clinical challenges. These include unraveling underlying physiological mechanisms, identifying pathogenic factors contributing to disorders, analyzing intricate biological samples, conducting drug screening experiments, discovering diagnostic biomarkers, and elucidating the mechanisms behind drug action and resistance.

Nemes et al. [31] proposed a detailed protocol for investigating and quantifying metabolites in individual isolated neurons through the utilization of single-cell

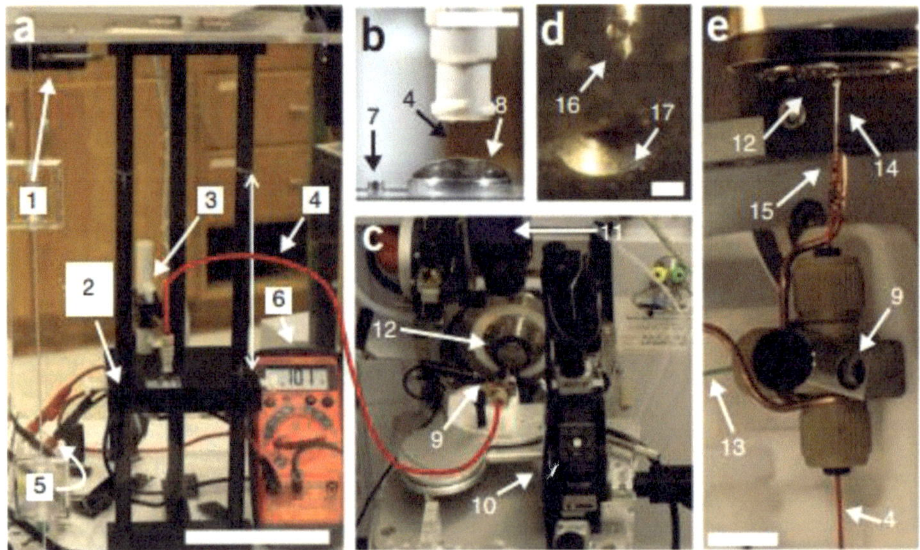

Figure 3.
Experimental setup of the single-cell CE-ESI-MS system. (a) the front view of the CE platform: (1) the enclosure equipped with a safety door; (2) platform for sample loading; (3) a holder that allows manual positioning of the separation capillary; (4) the separation capillary; (5) a resistor connected in series to a stable high-voltage power supply and the CE platform; (6) a digital multimeter connected in parallel to measure voltage drop on the resistor. The scale bar is 10 cm. (b) A magnified view of the sample-loading platform includes (7) a sample-loading vial and (8) an electrolyte-containing vial with the separation capillary positioned 2 mm below the electrolyte meniscus. The scale bar is 1 cm. (c) A distant view of the CE-ESI-MS ion source consists of (9) the CE-ESI interface equipped on (10) a three-axis translation stage; (11) CCD camera; and (12) inlet of mass spectrometer. (d) A magnified view shows (16) the stable Taylor cone (17) and the orifice of the mass spectrometer sampling plate. The scale bar is 500 µm. (e) Close-up view of the CE-ESI-MS ion source highlighting a T-union that houses fused silica capillaries for CE separation and (13) ESI sheath solution delivery, as well as (14) a metal emitter grounded through (15) a copper wire. The scale bar is 1 cm [31].

CE combined with electrospray ionization time-of-flight MS (**Figure 3**). The procedure necessitated approximately 2 hours for preparing the sample, isolating neurons, and extracting metabolites, and an additional hour for measuring metabolism. By utilizing this method, over 300 distinct compounds within the mass range of typical metabolites were detected in various individual neurons ranging from 25 to 500 µm in diameter. A subset of identified compounds was found to be adequate for revealing metabolic differences among freshly isolated neurons of different types as well as changes in the metabolite profiles of cultured neurons. Using the CE-MS approach combined with whole-cell patch clamp, Aerts et al. successfully detected 60 metabolites from only 3 pl of cytoplasm within a single cell. This technique allowed them to observe differences in the metabolome of heterogeneous cells within the brain by analyzing samples obtained from specific cellular structures [32].

Onjiko et al. [33] developed a CE-ESI-MS method to analyze metabolites in the genome expression of various cell types. They identified 40 metabolites from *Xenopus laevis* embryos at the 16-cell stage, which had distinct tissue fates. The findings indicated that certain metabolites exhibited varying levels of activity among wild-type and unperturbed embryos' cell types. Advancements also have been achieved by this research group in the analysis of embryo metabolism, which is a difficult task due to its diminishing cell sizes and intricate three-dimensional transformations.

They successfully conducted metabolic analysis on live frog embryos by incorporating capillary microsampling, micro-scale metabolite extraction, and CE-ESI-MS techniques, resulting in the identification of 230 distinct molecular characteristics. This approach has expanded the potential for acquiring more profound understanding of the process of embryonic development [34].

CE-MS also has been utilized in the metabolism analysis of plants and microbes. The physiological metabolism of these organisms is greatly influenced by environmental conditions, but there remains limited knowledge regarding the impact of such factors on their dynamic metabolic changes. To address this gap, a CE-MS approach was employed to analyze the transcript and metabolism of *Synechocystis* sp. PCC 6803 under nitrogen starvation [35]. The simultaneous analysis of 161 metabolites in *Synechocystis* sp. PCC 6803 was conducted before and after nitrogen depletion. The study revealed that the levels of glycogen and downstream metabolites involved in sugar breakdown increased during nitrogen starvation in *Synechocystis* sp. PCC 6803, thereby reconciling the discrepancy between upregulated sugar catabolic genes and glycogen accumulation observed in this unicellular cyanobacterium.

3.2 Proteins

The utilization of CE has greatly enhanced our understanding of the proteome in individual cells. Lombard-Banek et al. [36] conducted a pioneering quantitative proteomics investigation on blastomeres, identifying and quantifying 438 protein groups across three distinct types: D11, V11, and V21. Interestingly, it is noteworthy that the majority of proteins found in individual blastomeres of the same cell type (such as D11) exhibited a significant degree of overlap. However, there were a few instances where protein expression did not coincide, which could potentially indicate divergent developmental paths for the daughter cells originating from these specific blastomeres. Moreover, their research focus then shifted toward examining blastocyst cells

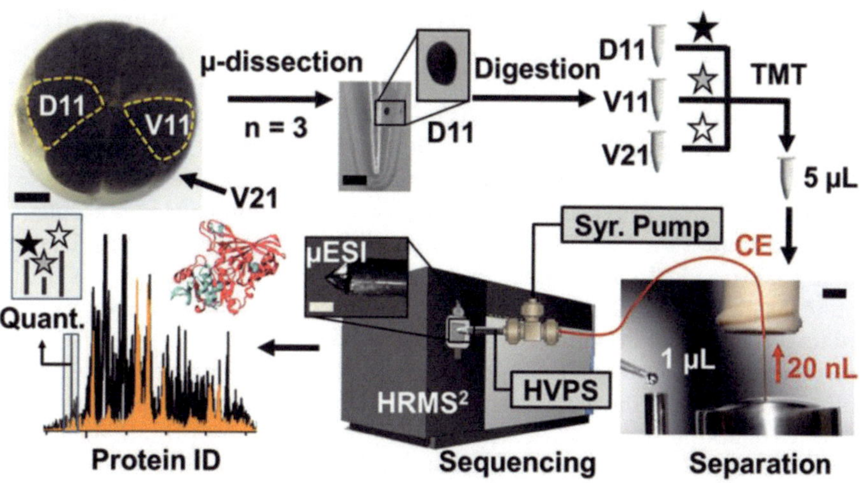

Figure 4.
Microanalytical process that allows for the simultaneous quantification of multiple proteins in individual cells within the 16-cell Xenopus embryo. This approach involves microdissection, micro-scale bottom-up proteomics, and a specially designed single-cell CE-μESI platform coupled with a high-resolution tandem mass spectrometer (HRMS²). HVPS: high-voltage power supply; Syr. Pump: syringe pump. Scale bars: 150 μm (embryo and μESI, left-middle panels), 250 μm (microcentrifuge vial), 1.5 mm (separation, right panel) [37].

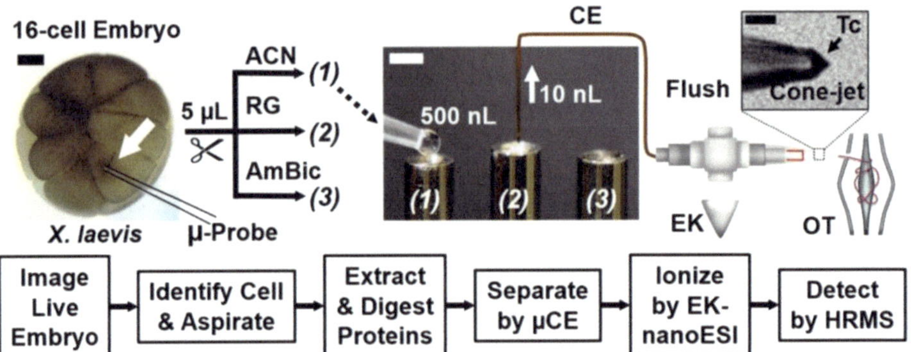

Figure 5.
Proteomic analysis of a single cell in a live X. laevis embryo without the use of labeling techniques. The given example demonstrates the optical recognition of the midline animal-dorsal cell (D11, indicated by a white arrow) in the 16-cell embryo. Approximately 20 nl of cellular content were extracted using capillary microsampling techniques employing a microprobe. The aspirated sample containing proteins was subjected to extraction and digestion processes. Subsequently, peptides were separated utilizing a microanalytical CE platform and ionized through a custom-designed electrokinetically pumped nanoelectrospray ionization (EK-nESI) source. This source was continuously monitored using a long-working distance camera to ensure stable operation within the cone-jet regime (referred to as Taylor cone or Tc), thereby facilitating efficient ionization. The resulting peptide ions were then analyzed utilizing an Orbitrap MS with high-resolution capabilities. Scale bars are provided for reference: 200 µm in black and 1 mm in white [38].

within the 16-cell *Xenopus* embryos (**Figure 4**) [37]. The single blastocyst cells were isolated through microdissection and underwent pre-treatment procedures, including cell lysis, protein reduction, alkylation, and overnight enzymatic hydrolysis. The proteins were then analyzed qualitatively and quantitatively using CE, microelectrospray ionization, and high-resolution mass spectrometry (CE-µESI-HRMS). The study achieved coverage of approximately 30% of the predicted proteome for individual *X. laevis* blastomeres, with the identification of 1630 proteins from only 20 ng of non-vitelline proteins in a single blastocyst cell measuring about 150 µm in diameter. **Figure 5** presented an alternative approach developed by this team, which involved the integration of subcellular capillary microsampling, one-step protein extraction and digestion, peptide separation using CE, ionization through electrokinetically pumped nanoelectrospray, and detection using high-resolution Orbitrap MS. This method enabled the direct analysis of proteomics in individual live cells, specifically in vertebrate embryos. This breakthrough represented a significant advancement as it overcame the limitation that MS was not compatible with complex tissues. The study eventually identified and quantified approximately 750–800 protein groups with exceptional sensitivity and accuracy in detection. Moreover, this technology demonstrated its scalability for application in live zebrafish embryos [38].

The recent advancements in CE-MS have significantly improved its detection capabilities by introducing a reversed-phase C18 microcolumn for pre-separation prior to CE, enabling the identification of 141 proteins from a protein digest sample as low as 500 pg and up to 737 proteins from a protein digest sample of just 1 ng [39]. And the PANC-1 cells (around 20 µm in diameter) have been among the tiniest cell types examined using CE. During analysis, researchers noted slight variations in the phosphorylation rates of protein kinase B among individual cells [40].

4. Application of electrophoresis in the ionization of the analytes from single cells for MS analysis

Ensuring that the protein states detected in single-cell MS are consistent with those within the cell is a crucial issue. Huang's research team postulated that the optimal accuracy can be achieved by promptly detecting proteins within a time frame of 100 ms after their release from the cell. Consequently, they developed an "in-cell" MS method for on-site analysis of proteins and protein complexes in living cells [41–43]. **Figure 6a** illustrates the integration of online electroporation-induced protein release and millisecond-scale micro-electrophoretic separation into a standard nanospray needle. A low-frequency pulsed high potential, generated in-house, is applied through a ring silver electrode mounted outside the needle to initiate online electroporation, fast separation, and nESI successively without any physical contact between the electrode and cells. In a typical experiment, living cells are injected into the needle via a syringe preloaded with cell suspensions. The cells pass through an area with high potential due to nanospray force and targeted proteins are released through online electroporation (**Figure 6b**). This procedure is crucial for extracting protein samples from cells without the need for conventional sampling techniques. Instead, protein extraction in "in-cell" MS occurs automatically and continuously, resembling electroporation when high potential is applied. Following this automated release-based protein extraction, ultrafast electrophoresis is implemented to address or mitigate any potential matrix effects that may arise from the simultaneous release

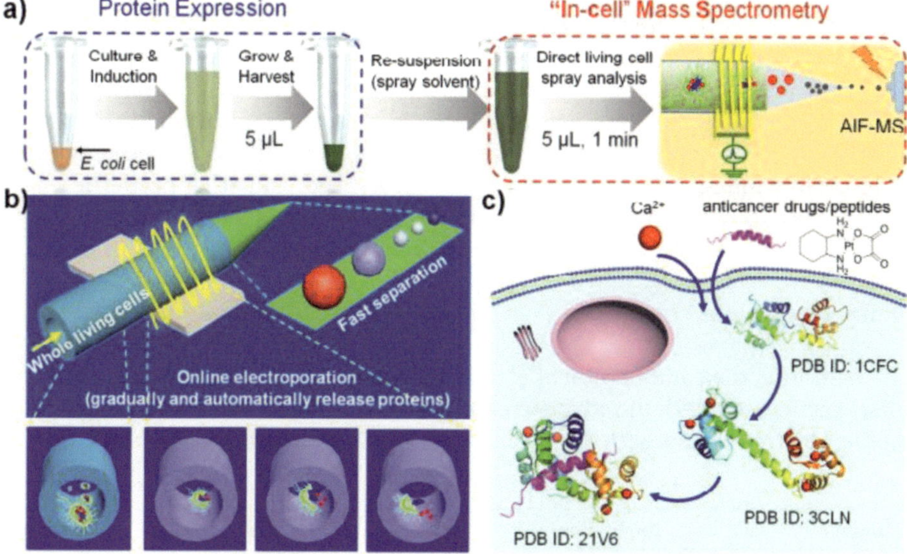

Figure 6.
Schematic diagram of (a) "in-cell" MS online electroporation, (b) following millisecond microelectrophoresis setups, and (c) its application for in-situ monitoring of dynamic protein-ligand. As demonstrated, the enhanced desolvation effectiveness in the all-ion-fragmentation mode allows for the identification of individual proteins and detection of larger protein complexes within living cells using "in-cell" MS. It is important to acknowledge that various factors, such as cell penetration efficiency, reaction rates, concentration, and intracellular constituents, influence the binding states between proteins and ligands. AIF-MS refers to all-ion-fragmentation MS [41].

of cellular components, such as inorganic salts and organic small molecules. It is important to mention that both the online electroporation and separation procedures can be completed within a very short time frame of milliseconds. Moreover, "in-cell" MS utilizes nESI as an ionization technique to promptly convert proteins in solution released from living cells into the gas phase, followed by high-resolution Orbitrap mass identification.

In conclusion, the unique features of "in-cell" MS encompass the avoidance of offline lysis and protein sampling, elimination of offline separation, and minimal manipulation processes on cells. These characteristics facilitate the characterization of larger proteins and protein complexes within intact living cells (**Figure 6c**), making high-throughput analysis achievable. The utilization of this approach has facilitated the detection of a broader range of protein complexes, including 17 proteins with molecular weights ranging from 4 to 44 kDa. Additionally, the "in-cell" MS technique allows for real-time monitoring of dynamic protein interactions in live cells, and its feasibility has been demonstrated through the test of the interaction between calcium-regulated calmodulin and melittin. This innovative method for protein analysis shows significant potential applications in various fields, such as biochemistry labs, protein engineering, and the protein industry.

5. Conclusions and future directions

5.1 Conclusions

As mentioned earlier, the incorporation of electrophoresis and MS into experimental workflows has greatly promoted the development of single-cell analysis. Electrophoresis plays an important role in all stages of single-cell MS analysis, including the sampling process, separation of analytes, and ionization (**Table 1**). By integrating with single-cell MS in various ways, electrophoresis can effectively enhance the accuracy, substance coverage, analysis speed, and other properties of single-cell MS, thus playing a crucial role in the fields of metabolomics, proteomics, metallomics, and beyond.

5.2 Future direction

In the future, the combination of electrophoresis and MS will become more diverse, particularly with the advancement of ambient ionization MS [44, 45]. Coupling electrophoresis with atmospheric pressure ionization sources will be a

Category	Application	References
Sampling	Suction of contents from single cells	[22]
	Extraction of analytes from single cells	[28]
Separation	Separation of the metabolites from single cells	[31–35]
	Separation of the proteins from single cells	[36–40]
Ionization	Ionization of the proteins in situ	[41–43]

Table 1.
Summary of the application of electrophoresis in the single-cell MS analysis.

crucial direction for development. Building upon this, the creation of stable electrophoresis-MS instruments is another important area to focus on as it can enhance the stability, reproducibility, automation, programming capabilities, analysis speed, and accuracy of the analytical process.

Acknowledgements

This work was supported by the National Natural Science Foundation of China (No. 21864001), and the Sub-project of Key Project at the Central Government Level: the ability establishment of sustainable use for valuable Chinese medicine resources (2060302-2101-01).

Conflict of interest

The authors declare no conflict of interest.

Author details

Hui Li and Jiaquan Xu*
Jiangxi Key Laboratory for Mass Spectrometry and Instrumentation, East China University of Technology, Nanchang, P.R. China

*Address all correspondence to: jiaquan_xu@foxmail.com

References

[1] Barabási AL, Oltvai ZN. Network biology: understanding the cell's functional organization. Nature Reviews Genetics. 2004;**5**:101-113. DOI: 10.1038/nrg1272

[2] Sun LL, Dubiak KM, Peuchen EH, Zhang ZB, Zhu GJ, Huber PW, et al. Single cell proteomics using frog (*Xenopus laevis*) blastomeres isolated from early stage embryos, which form a geometric progression in protein content. Analytical Chemistry. 2016;**88**(13):6653-6657. DOI: 10.1021/acs.analchem.6b01921

[3] Klepárník K, Foret F. Recent advances in the development of single cell analysis—a review. Analytica Chimica Acta. 2013;**800**:12-21. DOI: 10.1016/j.aca.2013.09.004

[4] Wang Q, Fan JW, Bian XY, Yao H, Yuan XH, Han Y, et al. A microenvironment sensitive pillar[5] arene-based fluorescent probe for cell imaging and drug delivery. Chinese Chemical Letters. 2022;**33**(4):1979-1982. DOI: 10.1016/j.cclet.2021.10.040

[5] Sheng YW, Zhou MJ, You CJ, Dai XX. Dynamics and biological relevance of epigenetic N^6-methyladenine DNA modification in eukaryotic cells. Chinese Chemical Letters. 2022;**33**(5):2253-2258. DOI: 10.1016/j.cclet.2021.08.109

[6] Xu YZ, Zhang YF, Yang HH, Yin W, Zeng LL, Fang S, et al. Two-dimensional coordination polymer-based nanosensor for sensitive and reliable nucleic acids detection in living cells. Chinese Chemical Letters. 2022;**33**(2):968-972. DOI: 10.1016/j.cclet.2021.07.041

[7] Shen J, Chen J, Wang D, Liu ZJ, Han GM, Liu BH, et al. Real-time quantification of nuclear RNA export using an intracellular relocation probe. Chinese Chemical Letters. 2022;**33**(8):3865-3868. DOI: 10.1016/j.cclet.2021.10.032

[8] Yang XK, Zhang FL, Jin XK, Jiao YT, Zhang XW, Liu YL, et al. Nanoelectrochemistry reveals how soluble $A\beta_{42}$ oligomers alter vesicular storage and release of glutamate. Proceedings of the National Academy of Sciences of the United States of America. 2023;**120**(19):e2219994120. DOI: 10.1073/pnas.2219994120

[9] Qi YT, Jiang H, Wu WT, Zhang FL, Tian SY, Fan WT, et al. Homeostasis inside single activated phagolysosomes: quantitative and selective measurements of submillisecond dynamics of reactive oxygen and nitrogen species production with a nanoelectrochemical sensor. Journal of the American Chemical Society. 2022;**144**(22):9723-9733. DOI: 10.1021/jacs.2c01857

[10] Yang XK, Zhang FL, Wu WT, Tang Y, Yan J, Liu YL, et al. Quantitative nano-amperometric measurement of intravesicular glutamate content and of its sub-quantal release by living neurons. Angewandte Chemie International Edition. 2021;**60**(29):15803-15808. DOI: 10.1002/anie.202100882

[11] Jiang H, Qi YT, Wu WT, Wen MY, Liu YL, Huang WH. Intracellular monitoring of nadh release from mitochondria using a single functionalized nanowire electrode. Chemical Science. 2020;**11**:8771-8778. DOI: 10.1039/D0SC02787A

[12] Wei X, Yang M, Jiang Z, Liu JH, Zhang X, Chen ML, et al. A modular single-cell pipette microfluidic chip coupling to ETAAS and ICP MS for single

cell analysis. Chinese Chemical Letters. 2022;**33**(3):1373-1376. DOI: 10.1016/j.cclet.2021.08.024

[13] Sun CL, Ma CX, Li LL, Han YH, Wang DJ, Wan X. A novel on-tissue cycloaddition reagent for mass spectrometry imaging of lipid C=C position isomers in biological tissues. Chinese Chemical Letters. 2022;**33**(4):2073-2076. DOI: 10.1016/j.cclet.2021.08.034

[14] Zhang WM, Han ZS, Liang YQ, Zhang Q, Dou XN, Guo GS, et al. A pico-HPLC-LIF system for the amplification-free determination of multiple miRNAs in cells. Chinese Chemical Letters. 2021;**32**(7):2183-2186. DOI: 10.1016/j.cclet.2020.12.007

[15] Nahar L, Onder A, Sarker SD. A review on the recent advances in HPLC, UHPLC and UPLC analyses of naturally occurring cannabinoids (2010-2019). Phytochemical Analysis. 2020;**31**(4):413-457. DOI: 10.1002/pca.2906

[16] Zhang QF, Xiao HM, Zhan JT, Yuan BF, Feng YQ. Simultaneous determination of indole metabolites of tryptophan in rat feces by chemical labeling assisted liquid chromatography-tandem mass spectrometry. Chinese Chemical Letters. 2022;**33**(11):4746-4749. DOI: 10.1016/j.cclet.2022.01.004

[17] Harstad RK, Johnson AC, Weisenberger MM, Bowser MT. Capillary electrophoresis. Analytical Chemistry. 2016;**88**(1):299-319. DOI: 10.1021/acs.analchem.5b04125

[18] Voeten RLC, Ventouri IK, Haselberg R, Somsen GW. Capillary electrophoresis: trends and recent advances. Analytical Chemistry. 2018;**90**(3):1461-1481. DOI: 10.1021/acs.analchem.8b00015

[19] Mizuno H, Tsuyama N, Harada T, Masujima T. Live single-cell video-mass spectrometry for cellular and subcellular molecular detection and cell classification. Journal of Mass Spectrometry. 2008;**43**(12):1692-1700. DOI: 10.1002/jms.1460

[20] Esaki T, Masujima T. Fluorescence probing live single-cell mass spectrometry for direct analysis of organelle metabolism. Analytical Sciences. 2015;**31**(12):1211-1213. DOI: 10.2116/analsci.31.1211

[21] Zhang LW, Foreman DP, Grant PA, Shrestha B, Moody SA, Villiers F, et al. In situ metabolic analysis of single plant cells by capillary microsampling and electrospray ionization mass spectrometry with ion mobility separation. Analyst. 2014;**139**(20):5079-5085. DOI: 10.1039/C4AN01018C

[22] Yin RC, Prabhakaran V, Laskin J. Quantitative extraction and mass spectrometry analysis at a single-cell level. Analytical Chemistry. 2018;**90**(13):7937-7945. DOI: 10.1021/acs.analchem.8b00551

[23] Gong XY, Zhao YY, Cai SQ, Fu SJ, Yang CD, Zhang SC, et al. Single cell analysis with probe ESI-mass spectrometry: detection of metabolites at cellular and subcellular levels. Analytical Chemistry. 2014;**86**(8):3809-3816. DOI: 10.1021/ac500882e

[24] Yu Z, Chen LC, Ninomiya S, Mandal MK, Hiraoka K, Nonami H. Piezoelectric inkjet assisted rapid electrospray ionization mass spectrometric analysis of metabolites in plant single cells via a direct sampling probe. Analyst. 2014;**139**(22):5734-5739. DOI: 10.1039/C4AN01068J

[25] Deng JW, Yang YY, Xu MZ, Wang XW, Lin L, Yao ZP, et al. Surface-

coated probe nanoelectrospray ionization mass spectrometry for analysis of target compounds in individual small organisms. Analytical Chemistry. 2015;**87**(19):9923-9930. DOI: 10.1021/acs.analchem.5b03110

[26] Phelps MS, Verbeck GF. A lipidomics demonstration of the importance of single cell analysis. Analytical Methods. 2015;**7**(9):3668-3670. DOI: 10.1039/C5AY00379B

[27] Sun M, Yang ZB. Metabolomic studies of live single cancer stem cells using mass spectrometry. Analytical Chemistry. 2019;**91**(3):2384-2391. DOI: 10.1021/acs.analchem.8b05166

[28] Song LL, Chingin K, Wang M, Zhong DC, Chen HW, Xu JQ. Polarity-specific profiling of metabolites in single cells by probe electrophoresis mass spectrometry. Analytical Chemistry. 2022;**94**(10):4175-4182. DOI: 10.1021/acs.analchem.1c03997

[29] DeLaney K, Sauer CS, Vu NQ, Li LJ. Recent advances and new perspectives in capillary electrophoresis-mass spectrometry for single cell "omics". Molecules. 2019;**24**(1):42. DOI: 10.3390/molecules24010042

[30] Zhou W, Zhang BF, Liu YK, Wang CL, Sun WQ, Li W, et al. Advances in capillary electrophoresis-mass spectrometry for cell analysis. Trends in Analytical Chemistry. 2019;**117**:316-330. DOI: 10.1016/j.trac.2019.05.011

[31] Nemes P, Rubakhin SS, Aerts JT, Sweedler JV. Qualitative and quantitative metabolomic investigation of single neurons by capillary electrophoresis electrospray ionization mass spectrometry. Nature Protocols. 2013;**8**(4):783-799. DOI: 10.1038/nprot.2013.035

[32] Aerts JT, Louis KR, Crandall SR, Govindaiah G, Cox CL, Sweedler JV. Patch clamp electrophysiology and capillary electrophoresis–mass spectrometry metabolomics for single cell characterization. Analytical Chemistry. 2014;**86**(6):3203-3208. DOI: 10.1021/ac500168d

[33] Onjiko RM, Moody SA, Nemes P. Single-cell mass spectrometry reveals small molecules that affect cell fates in the 16-cell embryo. Proceedings of the National Academy of Sciences of the United States of America. 2015;**112**(21):6545-6550. DOI: 10.1073/pnas.1423682112

[34] Onjiko RM, Portero EP, Moody SA, Nemes P. In situ microprobe single-cell capillary electrophoresis mass spectrometry: metabolic reorganization in single differentiating cells in the live vertebrate (*Xenopus laevis*) embryo. Analytical Chemistry. 2017;**89**(13):7069-7076. DOI: 10.1021/acs.analchem.7b00880

[35] Osanai T, Oikawa A, Shirai T, Kuwahara A, Iijima H, Tanaka K, et al. Capillary electrophoresis–mass spectrometry reveals the distribution of carbon metabolites during nitrogen starvation in *Synechocystis* sp. PCC 6803. Environmental Microbiology. 2014;**16**(2):512-524. DOI: 10.1111/1462-2920.12170

[36] Lombard-Banek C, Reddy S, Moody SA, Nemes P. Label-free quantification of proteins in single embryonic cells with neural fate in the cleavage-stage frog (*Xenopus laevis*) embryo using capillary electrophoresis electrospray ionization high-resolution mass spectrometry (CE-ESI-HRMS). Molecular & Cellular Proteomics. 2016;**15**(8):2756-2768. DOI: 10.1074/mcp.M115.057760

[37] Lombard-Banek C, Moody SA, Nemes P. Single-cell mass spectrometry for discovery proteomics: quantifying translational cell heterogeneity in the 16-cell frog (*Xenopus*) embryo. Angewandte Chemie International Edition. 2016;**55**(7):2454-2458. DOI: 10.1002/anie.201510411

[38] Lombard-Banek C, Moody SA, Manzini MC, Nemes P. Microsampling capillary electrophoresis mass spectrometry enables single-cell proteomics in complex tissues: developing cell clones in live *Xenopus laevis* and zebrafish embryos. Analytical Chemistry. 2019;**91**(7):4797-4805. DOI: 10.1021/acs.analchem.9b00345

[39] Choi SB, Lombard-Banek C, Muñoz-LLancao P, Manzini MC, Nemes P. Enhanced peptide detection toward single-neuron proteomics by reversed-phase fractionation capillary electrophoresis mass spectrometry. Journal of the American Society for Mass Spectrometry. 2018;**29**(5):913-922. DOI: 10.1007/s13361-017-1838-1

[40] Mainz ER, Wang QZ, Lawrence DS, Allbritton NL. An integrated chemical cytometry method: shining a light on akt activity in single cells. Angewandte Chemie International Edition. 2016;**55**(42):13095-13098. DOI: 10.1002/anie.201606914

[41] Li GY, Yuan S, Zheng SH, Liu YZ, Huang GM. In situ living cell protein analysis by single-step mass spectrometry. Analytical Chemistry. 2018;**90**(5):3409-3415. DOI: 10.1021/acs.analchem.7b05055

[42] Li GY, Yuan S, Pan Y, Liu YZ, Huang GM. Binding states of protein–metal complexes in cells. Analytical Chemistry. 2016;**88**(22):10860-10866. DOI: 10.1021/acs.analchem.6b00032

[43] Chen YT, Li GY, Yuan SM, Pan Y, Liu YZ, Huang GM. Ultrafast microelectrophoresis: behind direct mass spectrometry measurements of proteins and metabolites in living cell/cells. Analytical Chemistry. 2019;**91**(16):10441-10447. DOI: 10.1021/acs.analchem.9b00716

[44] Feider CL, Krieger A, DeHoog RJ, Eberlin LS. Ambient ionization mass spectrometry: recent developments and applications. Analytical Chemistry. 2019;**91**(7):4266-4290. DOI: 10.1021/acs.analchem.9b00807

[45] Zhang XL, Zhang H, Wang XC, Huang KK, Wang D, Chen HW. Advances in ambient ionization for mass spectrometry. Chinese Journal of Analytical Chemistry. 2018;**46**(11):1703-1713. DOI: 10.1016/S1872-2040(18)61122-3

Chapter 2

Advantages of Serum Bovine Blood Electrophoresis in Veterinary Diagnosis

Ahmed Khiredine Metref

Abstract

The veterinary practitioner, for the diagnosis of diseases, needs practical tools, fast, inexpensive, and, above all, available. The importance of this chapter lies in the reduction of various para-clinical examinations known in bovine medicine, which are often difficult to handle during a displacement in rural areas and which are also expensive. For this reason, serum protein electrophoresis (SPE) is an important component of laboratory diagnostic evaluations for serum protein measurement. Electrophoresis is based on the movement of charged particles through a buffered medium subjected to an electric field. Some variations that can be noticed in the SPE depend on some physiological and pathological cases. Early diagnosis of diseases is particularly important because treatments are no longer effective when the degree of consequences damages are too severe; because the clinical signs are not specific, the general clinical examination of the dairy cow can only lead to a suspicion of disease without a necessary tool, for confirmation or discover an insidious inflammatory process.

Keywords: electrophoresis, blood serum variations, bovine medicine, diagnosis, pathology

1. Introduction

Electrophoresis is the most widely used technique for standard fractionation of serum proteins in clinical biochemistry and molecular biology [1]. By this technique, serum proteins can be separated into albumin and globulin fractions by electrophoresis. This method consists of an electrical charge applied to agarose or cellulose gel matrix; electrophoresis capillary proteins are currently the most practiced, allowing serum proteins to migrate through the matrix and separate into bands based on their charge and size. The separation of protein fractions occurs in a free liquid medium created by the low-viscosity buffer, in which the application of high voltage generates an electro-osmotic flow, causing a rapid movement of proteins toward the cathode [2]. The matrix of migration is stained to detect the protein bands and read via densitometry to generate a tracing.

This allows a better separation of proteins with similar physicochemical characteristics, generating multiple subpeaks or narrower peaks [3]. Serum protein electrophoresis has been studied in animal and equine medicine, in particular for the clinical diagnosis of diseases characterized by dysproteinemia (leishmaniasis, ehrlichiosis, and feline infectious peritonitis) or to identify the presence of inflammation [4]. Electrophoresis is based on the movement of charged particles through a buffered medium subjected to an electric field [5]. Agarose gel is combined with an alkaline buffer solution (pH 9.1). A semi-automatic system usually does this. The separated proteins are stained with an "Amidoschwarz" solution, and the excess dye is removed in an acidic solution. The electrophoretic profiles are analyzed visually using a densitometer that gives an accurate relative quantification of each individualized area to detect anomalies. The densitometer reads the gel to define the relative concentrations (percentages) of each fraction. Agarose gel electrophoresis has several advantages over cellulose acetate (better reproducibility of results and greater clarity of electrophoretic strips). The migration takes place at 20 Watt (through two electrodes, one negative, the other positive), at 20°C (for about 7 minutes). The system performs all coloring, bleaching, and drying sequences. Then, we wait until the tank cools. The dry gel is removed for its final treatment: the reading will be done through a densitometer, which allows the relative concentrations (percentages) of each fraction to be defined. Serum proteins have a negative charge, so in an electric field, they migrate to the positive pole, and they are separated from each other by different bands depending on their size [6]. The speed of their movement depends on the characteristics of the isolated protein, including its electric charge, size, shape, strength of the electric field, type of medium used for separation, and temperature. After separation, the protein fractions are fixed in an acidic solution to denature the proteins and immobilize them on the support [7]. The proteins are then stained and quantified by density measurement, also providing graphical data so that they can be analyzed by computer depending on the electrophoretic system used [4].

2. The normal value of serum blood electrophoresis in cattle

In cattle, the normal profile of an SPE includes albumin and α-, β-, and γ-globulins (see **Figure 1**). The α-fraction is the most rapidly migrating protein relative to all

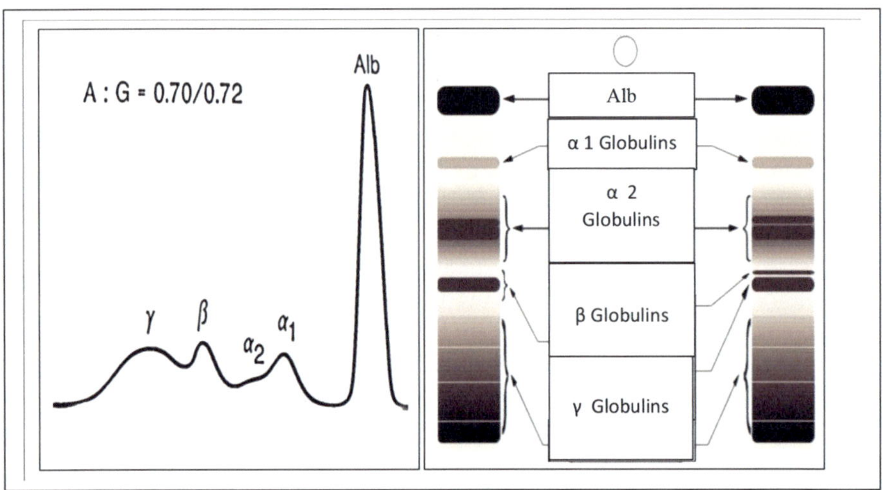

Figure 1.
Example of healthy bovine serum protein electrophoresis.

globulins, and in most species, it migrates as the α-fraction1 (fast) and α2 (slow). Many proteins of the acute phase of inflammation, considered diagnostically important, migrate in this fraction. Alpha1-antitrypsin, α1-acid glycoprotein, α1-antichymotrypsin, α1-ketoprotein, serum amyloid A, and α1-lipoprotein have been identified in the α1-globulin area, whereas haptoglobin, α2-microglobulin, α2-macroglobulin, ceruloplasmin, α2-antiplasmin, and α2-lipoprotein in the α2-globulin fraction [8]. This method (SPE) is also recommended to determine the distribution of globulins and quantify several fractions accurately (α-, β-, and γ-globulins), [9–11].

(A/G ratio ≈ 0.70/0.72) from [12].

3. The physiological variations of serum blood electrophoresis in cattle

The animal is considered clinically healthy if it does not show any decline in its zootechnical value, or expresses any sign of disease and influence of biological variations (the state of stress of an animal, the biological rhythm specific to each animal), the administration of drugs or tranquilizers [13]. We obtained a standard electrophoretic tracing, comprising five fractions: The first fraction is represented by albumin (Alb), followed by α1-, α2-, β-, and γ-globulins. In the physiological state, there is a clear dip between the end of the α2 fraction, and that of β-globulin (see **Figure 1**). This result corroborates with that found by authors [14–17]. Other authors describe a six-fraction separation using agarose gel electrophoresis: albumin, α1- and α2-, β1- and β2-, and γ-globulins [18]. The interpretation of the biochemical parameters measured can only be significant for an individual because each organism has a specific reactivity toward its environment, and a diagnostic opinion remains exclusively the domain of the clinician. The value of different components is described in **Table 1**.

3.1 Proteinogram variation in relation to age

The influence of age on proteinogram values has been studied by other authors and does not corroborate our results. Each author finds the age of the animals is one of the important factors that can affect the concentrations of the different serum protein fractions or their electrophoretic pattern, especially in the early months of life [20], and showed that in young and adult cattle, there is also the existence of an age-related influence for α- and γ-globulin fractions namely, the values of α1-globulins were higher in calves, while for adult animals they presented higher concentrations of γ-globulins [21].

Section	Value
Ratio: A/G	0.7
Albumin (Alb)	25–42 g/l
α1-globulins	6–12 g/l
α2-globulins	3.5–9.5 g/l
B-globulins	4–11 g/l
Γ-globulins	7–26 g/l

Table 1.
The normal value of different sections in bovine serum blood electrophoresis [19].

3.2 Proteinogram variation in relation to body condition score

The body condition of the animals is one of the indicators of the efficiency and safety of a ration, so for this reason, they directly impact the different components of SPE, especially the negative marker like albumin [16].

3.3 Variation of the proteinogram in relation to the physiological stage

The body condition varies significantly according to the physiological stage [22]; hence, the interest in the conduct of the dry period, which is a strategic and determining period for the nutritional future of the animal and the herd, his impact on SPE is significant by nutrition condition [17].

4. The pathological variations of serum blood electrophoresis in cattle

4.1 Albumin to globulin ratio (A/G)

Albumin is a small protein with a molecular weight of 69 kDa. The main functions of albumin are the maintenance of homeostasis by regulating oncotic pressure, blood pH buffering, and the transport of substances (fatty acids, fat-soluble hormones, and unconjugated bilirubin). It also acts as a free radical captor [23]. The liver is the principal site of albumin synthesis. Its degradation takes place in the liver, but other tissues may be involved: muscles, kidneys, and skin. The plasma concentration of albumin is determined by the intensity of hepatic synthesis, which is generally in balance with its elimination. It is responsible for about 75% of the plasma osmotic pressure and is a major source of amino acids that can be used by the animal's body when needed [24]. *Hypoalbuminemia* may therefore be caused by a lack of synthesis, which may be an indication of severe hepatic disorders. However, hypoalbuminemia may result from renal leakage caused by glomerulopathy. It may also be the result of severe inflammation of the intestine, leading to protein loss. Hypo-albuminemia is therefore not specific to liver disease. It is also not very sensitive since it only appears at the end of the evolution of a hepatic disease [8].

However, it is a consequence of poor rationing associated or not with parasitic infestation. This confirms that it is a good negative marker of inflammation, as previously established [8, 25]. The normal A/G ratio is between 0.6 and 0.9 in cows, but the relative concentrations of albumin and globulins can be changed in many conditions or diseases, resulting in changes in their ratio. Also, considers that in chronic and severe liver diseases, there is usually an increase in immunoglobulins (gamma-globulins: IgM, IgG, and IgA) together with a decrease in serum albumin concentration [25]. There may also be an increase in inflammatory proteins such as α2 and β-globulins. In the case of evolving abscesses, an increase in fibrinogen may be noted along with a decrease in the A/G ratio. Unfortunately, the decrease in the A/G ratio is not specific to liver damage: it can fall in many infectious or autoimmune diseases [26, 27], for that only serum protein electrophoresis (SPE) is of real interest in interpreting a drop in the A/G ratio (see **Figure 2**).

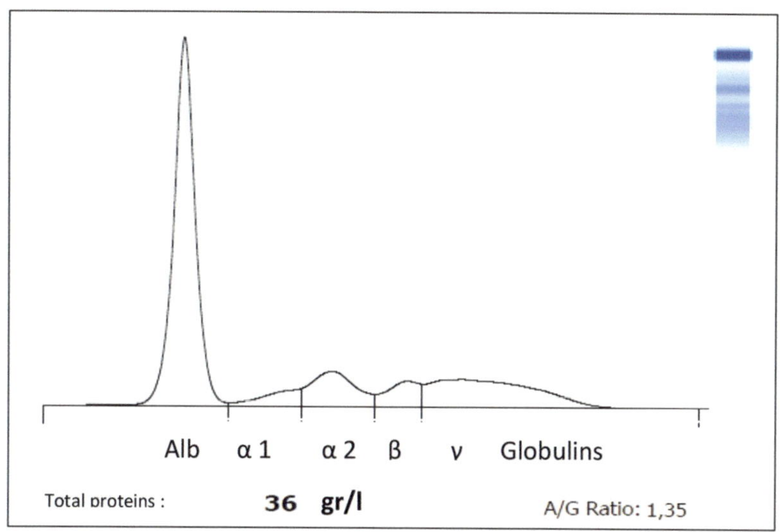

Figure 2.
Electrophoresis pattern of a 6-year-old cow with hypo-proteinemia [28].

4.2 α1-globulins

The alpha 1 zone is constituted by alpha 1 antitripsin, orosomucoid, and alpha 1 antichymotripsin. The usual value is 6 gr/l [19]. Most of the cattle present a low value [28]. This abnormality in comparison with normal range [19], we had a low wave, without specific clinical manifestations. This anomaly reported in humans, has not been previously revealed in ruminants. The studies at this time has not been able to determine the cause of these cases. We suggested the hypothesis of a congenital origin of α1 antitrypsin deficiency [the major protein of α1 zone] [8], like for humans, because of the similarities of the role and origin in the two species. However, in cattle, the serum levels of α1-antitripsin can be so low that their systemic detection is difficult [29]. This case is described in humans [30], also, it is the result of an impairment of the respiratory system [31–34]. However, only the CSF (cerebrospinal fluid) in cattle contained α1-antitripsin, with detectable value [29].

4.3 α2-globulins

The alpha 2 zone is constituted by haptoglobin, ceruloplasmin, Gc globulin, alpha 2 macroglobulin, and alpha-lipoproteins. We have a presence of peaks for the cases that present bronchopneumonia with a state of thinness [28]. The latter is suspected of developing chronic inflammatory conditions without clinical expression that would prevent them from gaining weight. It was noted that this coincides with an increase in total protein levels and acute phase proteins [35]. The α2 fraction obtained by electrophoresis is always increased during an inflammatory or infectious process since the majority of acute phase proteins (APPs) migrate to this area. It would therefore be wise to take this fact into account rather than using separate APP assays that require more expensive techniques. Alpha2-globulins are most often increased during bronchopneumonia [8].

4.4 β-globulins

They are composed of fractions: transferin, heopexin, betalipoprotein, and complement C3. This fraction has recorded an increase in its area in liver diseases [19, 36] because it is associated with an increase of specific enzymes to the liver function and total proteinemia (including the gamma-globulin fraction, alpha2globulin) with a pattern characterized by the formation of a block (β-γ) when the chronic development (see **Figure 3**). The cases of hypo-beta-globulins (small wave) appeared simultaneously with the cases of hypo-albuminemia, and this was in the absence of an increase of the enzymes of the hepatic function and without particular clinical signs. In these cases, we believe that there is a lack of food intake in relation to the physiological state and real needs of these animals [36]. alteration and bronchopneumonia [28].

4.5 γ-Globulins

They constitute the group of immunoglobulins (IgG, IgA, IgD, IgE, and IgM): we noticed that their increase is permanent during clinical conditions such as lameness, bronchopneumonia, and mastitis [28]. They could not be revealed by the other biochemical parameters assayed independently of the clinic; this makes them a good indicator of inflammation. We distinguish two representative groups of hyper-gammas-globulins: Those with monoclonal peaks (narrow and homogeneous): monoclonal gammopathy (MG) is characterized by the increase of a single type of Ig belonging to a well-defined class and subclass [37]. It is a synthesis of antibodies directed against possible specific antigens caused by viral, bacterial, or parasitic attacks [20]. As a result, the poor weight status of these cases of cattle, which do not express clinical manifestations, is explained [28]. Other authors have described these monoclonal peaks in this species for older subjects (exceeding 6 months) for lymphocytic leukemia or reticuloendothelial system tumors [8, 12]. Monoclonal gammopathy is characterized by a sharp, homogeneous, spike-like peak in the focal region of the γ-globulin area (see **Figure 4**).

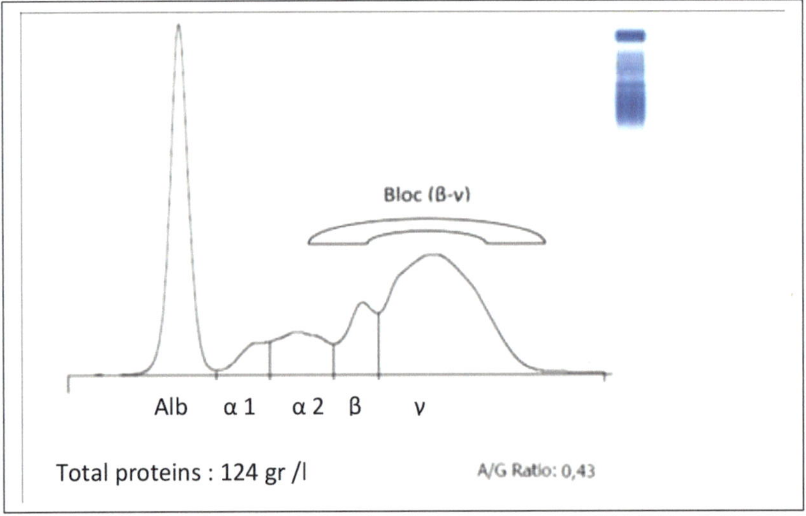

Figure 3.
Plot of blood serum electrophoresis of cow with liver.

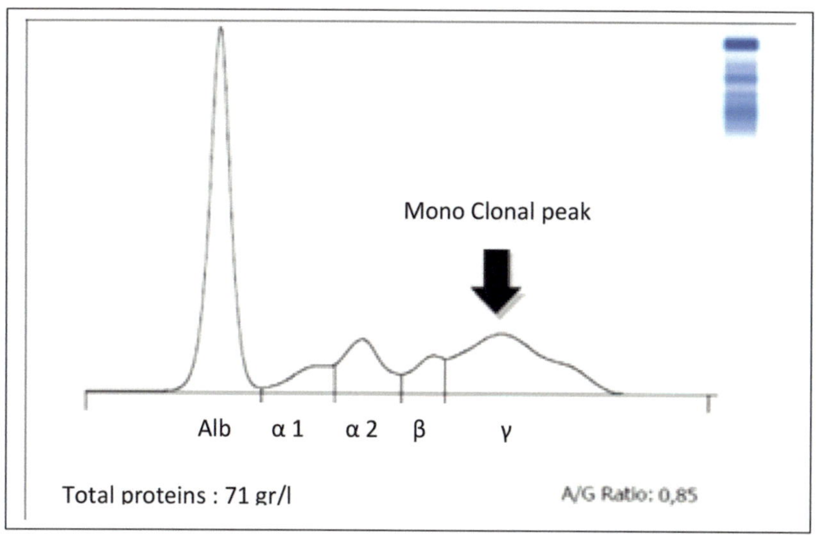

Figure 4.
Plot of blood serum electrophoresis of a young bull, 6 months old, delayed growth (monoclonal peak) [28].

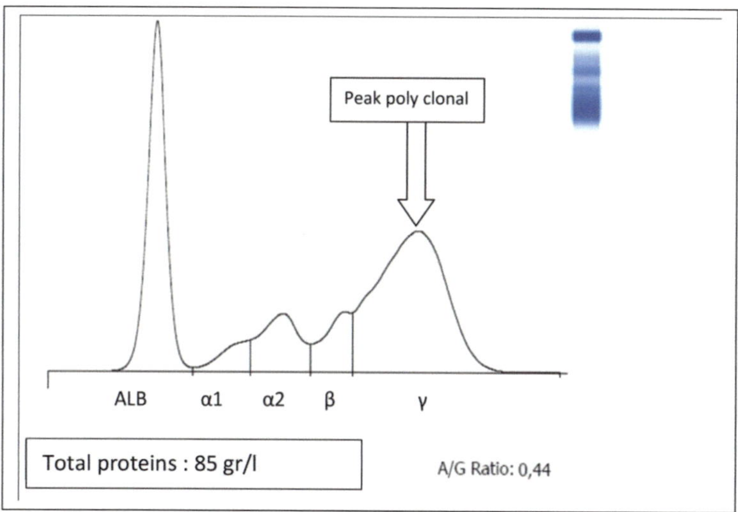

Figure 5.
Plot of blood serum electrophoresis of cow with chronic mastitis (poly clonal peak) [28].

The second group includes polyclonal hyper-gamma-globulins (diffuse increase). This is the case of cattle that develop general or organ-specific inflammatory diseases (see **Figure 5**) [38].

5. Conclusion

Electrophoresis alone can provide an answer to many clinical diagnostic problems. Its interpretation, in combination with the data collected for each individual, has made it possible to confirm clinically apparent cases and to discover other asymptomatic cases. We can consider serum protein electrophoresis as a practical tool for diagnosis in rural medicine: the ease of its realization, the simplicity of its method, the rapidity of its response (less than 1 h), and the quality of its information. The interpretation of the biochemical parameters measured can only be meaningful for an individual (a diagnostic choice remains exclusively the domain of the clinician). The study of the tracings allowed us to discover that electrophoresis alone can provide an answer to a large number of clinical diagnostic problems. Its interpretation in combination, with all the data collected for each individual, allowed us to confirm clinically apparent cases and to discover other asymptomatic cases. Electrophoresis should be combined with the clinical examination and other complementary examinations (Biopsy, Hematology, Bacteriology). It should be associated with other more specific biochemical assays (specific functional exploration of organs) in order to avoid the veterinary practitioner having to recommend uncertain treatments. The abnormal electrophoretic profile of serum proteins may be characteristic of certain disorders or diseases but, in other cases, may indicate only nonspecific pathological processes. Despite this low specificity in the diagnosis of certain diseases, the determination of serum protein profile in ruminants and the correct interpretation of their results are very useful for clinicians in the diagnosis of healthy and diseased animals and can serve as a basis for other specific laboratory examinations. We can consider serum protein electrophoresis as a practical tool for diagnosis in rural medicine: the ease of realization, the simplicity of its method, the rapidity of its response (less than 1 h), the quality of its information; electrophoresis constitutes, therefore, a practical and economical alternative in the semiology exploration of ruminant's medicine.

Conflict of interest

There is no conflict of interest. We do not benefit from any financial support from an organization or institution.

Author details

Ahmed Khiredine Metref
Clinic Department, Veterinary Institute, University in Ouled Yaïch, Algeria

*Address all correspondence to: metref_ahmed@univ-blida.dz

References

[1] O'Connell T, Horite T, Kasravi B. Understanding and interpreting serum protein electrophoresis. American Family Physician. 2005;**2005**(71): 105-112

[2] Giordano A, Paltrinieri S. Interpretation of capillary zone electrophoresis compared with cellulose acetate and agarose gel electrophoresis: Reference intervals and diagnostic efficiency in dogs and cats. Veterinary Clinical Pathology. 2010;**2010**(39):464-473

[3] Petersen Jr OA, Mohammad A, Da P. Capillary electrophoresis and its application in the clinical laboratory. Clinica Chimica Acta. 2003;**2003**(330):1-30

[4] Vavricka S, Burri E, Beglinger C, Degen L, Manz M. Serum protein electrophoresis: An underused but very useful test. Digestion. 2009;**2009**(79):203-210

[5] Azim W, Azim S, Ahmed K, Shafi H, Rafi T, Luqman M. Diagnostic significance of serum protein electrophoresis. Biomédica. 2004;**2004**(20):40-44

[6] Esmaeilnejad B, Tavassoli M, Asri-Rezaei S, Dalir-Naghadeh B, Pourseyed S. Evaluation of serum total protein concentration and protein fractions in sheep naturally infected with Babesia ovis. Comparative Clinical Pathology. 2014;**2014**(23):151-155

[7] Tymchak L. 2010. Amino acids and proteins. In: Bishop ML, Fody EP, Schoeff LE, editors. Clinical Chemistry: Techniques, Principles, Correlations. 6th ed. Philadelphia, USA: Lippincott Williams & Wilkins; 2010. pp. 223-265

[8] Smith B, van Metre D, Pusterla N. Book, Large Animal Internal Medicine. 6th ed. Elsevier Inc.; 2021. pp. 435-442

[9] Cray C, Rodriguez M, Zaias J. Protein electrophoresis in psittacine plasma. Veterinary Clinical Pathology. 2007;**2007**(36):64-72

[10] Cray C, Tatum LM. Applications of protein electrophoresis in avian diagnostics. Journal of Avian Medicine and Surgery. 1998;**1998**(12):4-10

[11] Metref A, Melizi M. Importance of serum protein electrophoresis in bovine medicine (easy tool for diagnosis). Global Veterinaria. 2017;**19**(2):509-516

[12] Kaneko JJ. A century of animal clinical biochemistry: growth, maturity and visions for the future. Revue de Medecine Veterinaire. 2000;**151**(7):601-605

[13] Stockham SL, Scott MA. Introductory concepts. In: Fundamentals of Veterinary Clinical Pathology. Blackwell Publishing Company, Iowa State Press; 2002. pp. 3-31

[14] Dartois HR. Contribution à la mise en œuvre d'une méthode d'analyse des protéines sériques par électrophorèse en gel d'agarose. Thèse. 2011. Ecole vétérinaire d'Alfort

[15] Errico G, Giordano A, Paltrinieri S. Diagnostic accuracy of electrophoretic analysis of native or defibrinated plasma using serum as a reference sample. Veterinary Clinical Pathology. 2012;**2012**(41):529-540

[16] Alberghina D, Giannetto C, Vazzana I, Ferrantelli V, Piccione G. Reference intervals for total protein concentration, serum protein fractions,

and albumin/globulin ratios in clinically healthy dairy cows. Journal of Veterinary Diagnostic Investigation. 2011;**2011**(23):111-114

[17] Piccione G, Messina V, Alberghina D, Giannetto C, Caselle S, Assenza A. Seasonal vari-ations in serum protein fractions of dairy cows during different physiological phases. Comparative Clinical Pathology. 2012;**2012**(21):1439-1443

[18] Nagy O, Tóthová C, Nagyová V, Kováč G. Comparison of serum protein electropho-retic pattern in cows and small ruminants. Acta Veterinaria Brno. 2015;**2015**(84):187-195

[19] Fontaine M. Book, Vade-mecum du vétérinaire, 15ème édition, office nationale des publications universitaire. 1993

[20] Tóthová C, Nagy O, Nagyová V, Kováč G. The concentrations of selected blood serum proteins in calves during the first three months of life. Acta Veterinaria Brno. 2016;**2016**(85):33-40

[21] Tóthová C, Nagy O, Seidel H, Kováč G. Serum protein electrophoretic pattern in clini-cally healthy calves and cows determined by agarose gel electrophoresis. Comparative Clinical Pathology. 2013;**2013**(22):15-20

[22] Drame ED, Hanzen C, Houtain JY, Laurent Y, Fall A. Profil de l'état corporel au cours du post-partum chez la vache laitière. Annales De Medecine Veterinaire. 1999;**1999**(143):265-270

[23] Hankins J. The role of albumin in fluid and electrolyte balance. Journal of Infusion Nursing. 2006;**2006**(29):260-265

[24] Mackiewicz A. Acute phase proteins and transformed cells. International Review of Cytology. 1997;**197**(170):225-300

[25] Stockham SL, Scott MA. Liver function. In: Fundamentals of Veterinary Clinical Pathology. Blackwell Publishing Company, Iowa State Press; 2002. pp. 461-487

[26] Js L. Albumin for end-stage liver disease. Korean Journal of Internal Medicine. 2012;**2012**(27):13-19

[27] Kelly R. Ruminant liver disease. In: Cotton M, editors. Gross Pathology of Ruminants. Proceedings Sydney: Post-Graduate Foundation in Veterinary Science. University of Sydney; 2003. pp. 81-96

[28] Metref A, Aouane N, Djamila K, Mohamed M. Abnormalities of serum protein electrophoresis in cattle, compare between assays for markers of inflammation and markers of liver alteration. International Journal of Veterinary Science. 2021;**10**(3):208-213

[29] Di Filippo PA, Lannes ST, Meireles MAD, Nogueira AFS, Ribeiro LMF, Graça FAS, et al. Acute phase proteins in serum and cerebrospinal fluid in healthy cattle: possible use for assessment of neurological diseases. Brazilian Journal of Veterinary Research. 2018;**38**:779-784

[30] Denden S, Lakhdar R, Leban N, Daimi H, Elhayek D, Knani J, et al. Development: Alpha 1 antitrypsin deficiency. Mediterranean Journal of Human Genetics. 2010;**1**:26-33

[31] Janciauskiene SM, Bals R, Koczulla R, Vogelmeier C, Köhnlein T, Welte T. The discovery of α1-antitrypsin and its role in health and disease. Respiratory Medicine. 2011;**105**:1129-1139

[32] Tuder RM, Janciauskiene SM, Petrache I. Lung disease associated with α1-antitrypsin deficiency. Proceedings of the American Thoracic Society. 2010;**7**:381-386

[33] Fagliari JJ, Passipieri M, Okuda HQ, Silva SL, Silva PC. Serum protein concentrations, including acute phase proteins, in calves with hepatogenous photosensitization. Arquivo Brasileiro de Medicina Veterinária e Zootecnia. 2007;**59**:1355-1358

[34] Sampaio PH, Fidelis JOL, Marques LC, Machado RZ, Barnabé PA, André MR, et al. Acutephase protein behavior in dairy cattle herd naturally infected with Trypanosoma vivax. Veterinary Parasitology. 2015;**211**:141-145

[35] Humblet MF, Coghe J, Lekeux P, Godeau JM. Acute phase proteins assessment for an early selection of treatments in growing calves suffering from bronchopneumonia under field conditions. Research in Veterinary Science. 2004;**2004**(77):41-47

[36] Camus M, Krimer P, Leroy B, Almy F. Evaluation of the positive predictive value of serum protein electrophoresis beta-gamma bridging for hepatic disease in three domestic animal species. Veterinary Pathology. 2010;**2010**(47):1064-1070

[37] Lecarrer D. Serum Protein Electrophoresis Immunofixation. SEBIA (INC); 1994. 125 p (11 p-31 p)

[38] Jackson D, Elsawa S. Factors regulating immunoglobulin production by normal and disease-associated plasma cells. Biomolecules. 2015;**5**:20-40

Chapter 3

Evaluation of Dental Materials and Oral Disease-Related Proteins in Dentistry: Efficacy of Electrophoresis as a Valuable Tool

Aida Meto and Agron Meto

Abstract

Electrophoresis is a versatile technique that allows for the separation of molecules based on their size and electrical charge. In the field of dentistry, electrophoresis is widely used in various applications, including the analysis of dental materials and proteins associated with diseases of the oral cavity. Through electrophoresis, it is possible to evaluate the size and distribution of filler particles within resin matrices, providing valuable information on the mechanical properties and durability of composite materials used in dental restorations. Furthermore, this technique has significantly contributed to the study of proteins implicated in oral diseases, such as dental caries and periodontitis. By effectively identifying and separating these proteins, researchers gain a deeper understanding of the mechanisms underlying these conditions, facilitating the development of innovative therapeutic strategies. Overall, the application of electrophoresis in dentistry has emerged as an indispensable tool for comprehensive analysis of dental materials and characterization of proteins associated with oral diseases.

Keywords: dental materials, electrophoresis in dentistry, oral diseases, proteins, biomarkers

1. Introduction

Shortly, electrophoresis is a powerful analytical technique widely used in dental research for the separation and analysis of various biomolecules [1]. In this chapter, we will mention how electrophoresis is employed to study proteins, nucleic acids, glycosylated proteins, phosphorylated proteins, and other relevant molecules in dental materials and oral health research. Understanding the composition, modifications, and interactions of these biomolecules is critical for unraveling the molecular mechanisms underlying dental diseases, identifying potential biomarkers, and improving dental materials' performance.

1.1 Overview of electrophoresis

Electrophoresis is a widely used technique in dentistry research for the separation and analysis of dental materials and oral disease-related proteins [1, 2]. It is based on the principle of applying an electric field to migrate charged molecules in a medium, such as a gel or a capillary, according to their size, charge, or isoelectric point. Electrophoresis allows the separation, identification, and quantification of various components in complex mixtures, including proteins, nucleic acids, carbohydrates, and ions [3]. Different electrophoretic techniques are employed in dental research, depending on the specific objectives and requirements of the analysis. Gel electrophoresis, such as polyacrylamide gel electrophoresis (PAGE) and agarose gel electrophoresis, is commonly used for the separation of proteins and nucleic acids based on their size and charge [4]. Capillary electrophoresis (CE) offers high resolution and efficiency for the separation of small molecules and ions [5]. Two-dimensional electrophoresis (2DE) combines two separation dimensions, such as isoelectric focusing (IEF) and sodium dodecyl sulfate (SDS)-PAGE, enabling more comprehensive protein profiling [6]. Electrophoresis has revolutionized dental material analysis by allowing the assessment of material components, such as monomers, additives, and degradation products. It aids in the determination of molecular weight distribution, polymerization kinetics, and leaching characteristics of dental materials. Additionally, electrophoresis techniques have been instrumental in protein analysis, allowing the identification, characterization, and quantification of disease-related proteins in oral fluids, tissues, and biofluids [7].

1.2 Significance of dental material analysis

Dental material analysis plays a crucial role in dentistry research and clinical practice [8]. The selection and evaluation of dental materials are essential for the success and longevity of various dental treatments, including restorative procedures, prosthetic devices, and orthodontic appliances [8–10]. Dental materials must possess desirable physical, chemical, and biological properties to ensure their safety, biocompatibility, and effectiveness in clinical applications. Therefore, thorough analysis and characterization of dental materials are necessary to assess their quality, performance, and potential risks. By employing analytical techniques, researchers and clinicians can examine the composition, structure, and properties of dental materials [11]. This analysis helps in identifying any potential flaws, defects, or limitations in the materials, as well as understanding their behavior under different conditions. It allows for the development of improved dental materials with enhanced properties, such as increased strength, improved esthetics, and reduced toxicity [12]. Moreover, the analysis of dental materials aids in standardization, quality control, and regulatory compliance in the dental industry, as follows:

1.2.1 Analysis of proteins in dental research

Proteins play a pivotal role in dental research, serving as key structural components and orchestrating various biological processes. Electrophoresis techniques are commonly used to analyze proteins in dental materials and biological samples. Several classes of proteins are of particular interest [13–15]:

- Dental tissue proteins: Dental tissues, such as enamel, dentin, and cementum, have unique protein compositions. Electrophoresis can be used to study the protein profiles of these tissues, providing insights into their structural and functional properties. For example, amelogenins are essential enamel matrix proteins involved in enamel formation and can be analyzed using electrophoresis.

- Salivary proteins: Saliva contains a diverse array of proteins with various functions, including antimicrobial properties and lubrication of oral tissues. Electrophoresis can help identify and quantify specific salivary proteins associated with oral health and disease.

- Inflammatory proteins: Inflammatory responses are critical in periodontal disease and dental caries. Identifying and quantifying inflammatory proteins, such as cytokines and chemokines, in gingival crevicular fluid (GCF) or dental plaque can help elucidate the molecular basis of these diseases. Electrophoresis can be used to analyze the expression levels of these inflammatory proteins.

- Matrix metalloproteinases (MMPs): MMPs are involved in tissue remodeling and degradation of extracellular matrix components. Dysregulation of MMPs is associated with periodontal tissue destruction. Electrophoresis can be used to analyze MMP expression levels in diseased tissues.

1.2.2 Analysis of nucleic acids in dental research

Nucleic acids, including DNA and RNA, are essential biomolecules in dental research, providing information about genetic factors, microbial presence, and gene expression in oral health and disease. Electrophoresis is commonly used to study nucleic acids in dental research [16, 17]:

- Analysis of dental microbiome: Electrophoresis-based methods, such as polymerase chain reaction (PCR) and gel electrophoresis, can be used to study the oral microbiome by identifying specific bacterial DNA sequences associated with dental diseases. For instance, 16S rRNA gene sequencing can be performed on gel-separated PCR products to characterize the oral microbial community.

- Genetic analysis of dental conditions: Nucleic acid electrophoresis can aid in the identification of genetic variations associated with dental conditions such as *amelogenesis imperfecta* and *dentinogenesis imperfecta*. Genetic testing using gel electrophoresis can reveal specific mutations linked to these dental disorders.

1.2.3 Other biomolecules of interest in dental research

Beyond proteins and nucleic acids, other biomolecules are relevant in dental research. Electrophoresis can be applied to study [18, 19]:

- Glycosylated proteins: Glycosylation is an essential posttranslational modification that influences the functions of dental proteins. Lectin-based electrophoresis techniques can be employed to analyze glycosylated proteins in dental tissues. *For example*, lectin blotting can detect specific carbohydrate structures on proteins separated by gel electrophoresis.

- Phosphorylated proteins: Phosphorylation is crucial in signal transduction pathways in dental cells. Phospho-specific antibodies or Phos-tag SDS-PAGE can be used to detect and analyze phosphorylated proteins. Electrophoresis combined with immunoblotting allows the detection of phosphorylated protein bands.

1.3 Role of protein analysis in oral disease research

Proteins play a vital role in the pathogenesis, progression, and diagnosis of various oral diseases. Understanding the protein composition and alterations associated with oral diseases is essential for advancing our knowledge of their underlying mechanisms and developing effective diagnostic and therapeutic strategies [20]. Protein analysis provides valuable insights into disease-related biomarkers, protein-protein interactions, and signaling pathways involved in oral diseases. Oral diseases, such as periodontal diseases, dental caries, and oral cancers, involve complex molecular processes that impact the composition and expression of proteins. Biomarker discovery and validation are crucial for early detection, prognosis, and monitoring of oral diseases [21]. Proteomic analysis techniques allow the identification and quantification of disease-specific proteins, providing potential targets for intervention and personalized treatment approaches. Furthermore, protein analysis facilitates the evaluation of treatment outcomes, the assessment of disease progression, and the monitoring of therapeutic responses [22, 23].

1.4 Preparing samples for electrophoresis

Preparing samples for electrophoresis involves specific steps depending on the type of electrophoresis and the type of samples you are working with (e.g., DNA, RNA, proteins) [24, 25]. Here's a general guide on how to prepare samples for different types of electrophoresis:

1.4.1 DNA electrophoresis (agarose gel electrophoresis)

Agarose gel electrophoresis is commonly used to separate DNA fragments based on their size [24]. Here's how to prepare samples for this type of electrophoresis:

Materials needed:

- DNA samples
- DNA ladder (molecular weight markers)
- Agarose powder
- Tris acetate EDTA (TAE) and tris borate EDTA (TBE) buffer
- Ethidium bromide or a DNA-specific stain
- Gel loading buffer

Steps:

1. Sample extraction: Extract DNA from your biological material using appropriate DNA extraction methods.

2. Quantification: Measure the concentration of your DNA samples using a spectrophotometer or fluorometer.

3. DNA denaturation (if needed): For double-stranded DNA, denature the samples by heating them at around 95°C for a few minutes and then cooling them on ice. This step converts double-stranded DNA into single-stranded DNA.

4. Prepare the agarose gel: Mix agarose powder with TAE or TBE buffer and heat the mixture to dissolve the agarose. Pour the liquid agarose into a gel tray and insert a comb to create wells.

5. Prepare loading buffer: Mix your DNA samples with a gel loading buffer. The loading buffer will add density to your samples, helping them sink into the wells.

6. Load the samples: Carefully load the DNA samples and the DNA ladder (molecular weight markers) into the wells of the agarose gel using a micropipette.

7. Electrophoresis: Submerge the gel tray in an electrophoresis tank filled with TAE or TBE buffer. Apply an electric field and run the electrophoresis until the DNA bands have separated according to their size.

8. Visualization: Stain the DNA with ethidium bromide or a DNA-specific stain and visualize the separated bands under UV light.

1.4.2 Protein electrophoresis (SDS-PAGE)

SDS-PAGE is used to separate proteins based on their molecular weight [25]. Here's how to prepare protein samples for SDS-PAGE:

Materials needed:

- Protein samples
- Protein molecular weight markers
- SDS-PAGE running buffer
- SDS-PAGE sample buffer
- Reducing agent (e.g., β-mercaptoethanol or DTT)

Steps:

1. Sample extraction: Extract proteins from your biological material using appropriate protein extraction methods.

2. Protein quantification: Determine the protein concentration using methods, such as the Bradford assay or the BCA assay.

3. Protein denaturation and reducing: Mix your protein samples with an SDS-PAGE sample buffer containing a reducing agent (e.g., β-mercaptoethanol or DTT). The reducing agent helps to denature the proteins and break disulfide bonds.

4. Boiling: Heat the protein samples in the SDS-PAGE sample buffer at around 95°C for a few minutes to ensure complete denaturation.

5. Prepare the SDS-PAGE Gel: Assemble the SDS-PAGE gel according to the manufacturer's instructions or your lab's protocol.

6. Load the samples: Load the denatured protein samples and the protein molecular weight markers into the wells of the SDS-PAGE gel.

7. Electrophoresis: Submerge the gel in an electrophoresis tank filled with SDS-PAGE running buffer. Apply an electric field and run the electrophoresis until the protein bands have separated according to their size.

8. Visualization: Stain the proteins with Coomassie Brilliant Blue or other compatible protein stains, or use silver staining for more sensitive applications.

It's important to note that electrophoresis protocols can vary depending on the specific application and the type of samples being analyzed. Always follow established protocols and optimize the procedure for your particular experiment to ensure accurate and reliable results.

1.4.3 Separation and purification steps

Before performing electrophoresis, separation, and purification steps are often necessary to obtain high-quality samples and accurate results. The steps for separation and purification depend on the type of sample and the target molecules you are working with (e.g., DNA, RNA, proteins) [19, 24, 26]. Here are some common separation and purification methods used before electrophoresis:

DNA electrophoresis (agarose gel electrophoresis): For DNA electrophoresis, the following separation and purification steps are commonly performed:

1. DNA extraction: This is the initial step to isolate DNA from the biological material (e.g., cells, tissues, blood). Various extraction methods, such as phenol-chloroform extraction, column-based purification, or commercial DNA extraction kits, can be employed.

2. PCR or DNA digestion (optional): If you are working with specific DNA fragments, you may need to perform polymerase chain reaction (PCR) or DNA digestion (restriction enzyme digestion) to amplify or cut the DNA into the desired fragments.

3. DNA cleanup: After PCR or DNA digestion, it's common to perform a DNA cleanup step to remove any leftover primers, enzymes, or other impurities that could interfere with the electrophoresis results. DNA cleanup can be done using purification kits or by precipitation methods.

RNA electrophoresis: For RNA electrophoresis, the following steps are typically performed:

1. RNA extraction: RNA is extracted from the biological material (e.g., cells, tissues, bacteria, or viruses) using methods, such as phenol-chloroform extraction, column-based purification, or commercial RNA extraction kits.

2. RNA integrity check: Before proceeding with electrophoresis, it is essential to assess the integrity of the RNA. This can be done using techniques such as gel electrophoresis or capillary electrophoresis with RNA-specific dyes.

3. RNA cleanup: Similar to DNA cleanup, RNA cleanup is performed to remove impurities that might interfere with the electrophoresis results. Various purification methods can be used, such as column-based cleanup or precipitation methods.

Protein electrophoresis (SDS-PAGE): For protein electrophoresis, the following steps are typically performed:

1. Protein extraction: Proteins are extracted from the biological material using appropriate protein extraction methods, such as cell lysis, tissue homogenization, or subcellular fractionation.

2. Protein quantification: The concentration of proteins in the samples is determined using methods, such as the Bradford assay or the BCA assay.

3. Protein denaturation and reduction: Proteins are denatured and reduced using an SDS-PAGE sample buffer containing a reducing agent (e.g., β-mercaptoethanol or DTT). This step ensures that the proteins are in a linear, denatured state for electrophoresis.

4. Protein cleanup: Sometimes, a protein cleanup step is performed to remove unwanted contaminants or interfering substances that could affect the electrophoresis results. Cleanup methods can include precipitation, dialysis, or column-based purification.

These separation and purification steps are critical for obtaining pure and concentrated samples of DNA, RNA, or proteins, which will ultimately lead to successful and interpretable electrophoresis results. The choice of specific methods and techniques will depend on the experimental requirements and the quality of the starting biological material.

1.5 Preparation of samples for electrophoresis of dental materials

Preparing samples for electrophoresis of dental materials involves specific steps to ensure accurate and reliable results. The preparation process may vary depending on the type of dental material you are working with, such as dental composites, cements, or adhesives [1, 2, 8–12]. Here's a general outline of the sample preparation process:

1. Sample collection: Collect the dental material you want to analyze. This could be a cured dental composite, set dental cement, or a specific component of a dental adhesive. Ensure that the sample is representative of the material you want to study.

2. Homogenization or grinding: If the dental material is in a solid form, you may need to homogenize or grind it to create a homogeneous sample. This step is crucial for ensuring even distribution and consistent results during electrophoresis.

3. Solubilization or extraction: Depending on the type of dental material, you might need to extract or solubilize the relevant components. For instance, if you are working with dental composites or adhesives, you might need to extract the monomers or polymer matrix. Some dental cements may need to be dissolved to release the proteins or other components of interest.

4. Protein precipitation (if analyzing proteins): If analysis focuses on proteins present in the dental material, it may need to perform a protein precipitation step to concentrate the proteins and remove interfering substances. This step helps improve protein detection and separation during electrophoresis.

5. Protein denaturation and reduction (For SDS-PAGE): If you are using SDS-PAGE to analyze proteins, the samples need to be denatured and reduced before loading onto the gel. This step involves heating the samples in the presence of a denaturing agent (such as SDS) and a reducing agent (such as β-mercaptoethanol or DTT) to ensure that the proteins are in a linear, denatured state for electrophoresis.

6. Size separation (electrophoresis): Load the prepared samples onto the appropriate electrophoresis gel. The choice of gel (agarose or polyacrylamide) and the electrophoresis conditions (e.g., voltage, buffer system) will depend on the specific objectives of your study.

7. Staining or detection: After electrophoresis, you may need to stain the gel to visualize the separated components. Coomassie Brilliant Blue, silver staining, or specific protein stains are commonly used for protein detection. For other types of dental materials, alternative staining methods or detection techniques may be employed.

8. Data analysis and interpretation: Analyze the gel images to identify and quantify the components of interest. Compare the results with appropriate controls or standards to validate your findings.

It's essential to optimize the sample preparation process to obtain reliable and reproducible results. The specific steps and conditions will depend on the dental material you are studying and the objectives of your analysis. Additionally, some dental materials might require additional purification steps, depending on their complexity and the presence of interfering substances. Always follow established protocols and adapt them as needed for your specific research.

2. Electrophoresis principles and techniques

2.1 Gel electrophoresis

Gel electrophoresis is one of the most commonly used techniques in electrophoresis. It involves the migration of charged molecules through a gel matrix under the influence of an electric field. The gel matrix provides a medium for size-based separation of molecules, with smaller molecules migrating faster than larger ones [26].

2.1.1 Polyacrylamide gel electrophoresis

Polyacrylamide gel electrophoresis (PAGE) is widely used for protein separation. It utilizes polyacrylamide gels with different concentrations, which create sieving properties suitable for separating proteins based on their size. Two commonly used variants of PAGE are native PAGE and sodium dodecyl sulfate-polyacrylamide gel electrophoresis (SDS-PAGE). Native PAGE preserves the protein's native structure, while SDS-PAGE denatures proteins and allows for separation based on molecular weight [4, 6]. In dentistry, PAGE is utilized to analyze dental materials, such as composite resins and adhesive systems, to assess their polymerization efficiency, molecular weight distribution, and the presence of impurities or residual monomers. Additionally, PAGE is employed in protein analysis for the identification and characterization of disease-related proteins in oral tissues and fluids [27].

2.1.2 Agarose gel electrophoresis

Agarose gel electrophoresis is commonly used for the separation of nucleic acids, such as DNA and RNA. Agarose gels are formed by the polymerization of agarose, a polysaccharide derived from seaweed. The porosity of the gel matrix can be adjusted by varying the agarose concentration, allowing for the separation of nucleic acid fragments of different sizes [28]. While agarose gel electrophoresis is primarily employed for nucleic acid analysis, it can also be utilized in dental research for the examination of genetic factors related to oral diseases or the evaluation of antimicrobial agents targeting oral pathogens [29].

2.2 Capillary electrophoresis

Capillary electrophoresis (CE) is a high-resolution technique that utilizes a narrow capillary tube as the separation medium. The capillary is typically coated with a polymer or has a charged inner surface to facilitate the separation of analytes based on the charge-to-size ratio. CE offers advantages such as high separation efficiency, short analysis time, and small sample requirements [30]. In dentistry, CE is employed for the analysis of ions, small molecules, and biomarkers in oral fluids and dental materials. It allows for the quantification of metal ion release from dental alloys, the evaluation of antimicrobial agents, and the assessment of salivary biomarkers for oral diseases [31, 32].

2.3 Two-dimensional electrophoresis

Two-dimensional electrophoresis (2DE) is a powerful technique that combines two separation dimensions to achieve increased resolution and separation capacity [33]. The first dimension typically involves isoelectric focusing (IEF), which separates proteins based on their isoelectric points (IP). In the second dimension, proteins are separated based on their molecular weights using SDS-PAGE [34]. The 2DE technique enables the separation and visualization of complex protein mixtures, providing a comprehensive protein profile. In dentistry, 2DE is utilized to analyze oral disease-related proteins, identify biomarkers, and compare protein expression patterns in healthy and diseased oral tissues [35].

2.4 Isoelectric focusing

IEF is a technique that separates proteins based on their IP, which is the pH at which a protein has no net charge. IEF utilizes a pH gradient gel or a capillary with a pH gradient, and proteins migrate until they reach their isoelectric point and become focused in a narrow band [36]. IEF is often used as the first dimension in 2DE to achieve high-resolution separation based on IP. In dentistry research, IEF is employed to study protein charge heterogeneity, identify protein isoforms, and analyze protein modifications [37].

3. Electrophoresis analysis of dental materials

In dentistry, the degradation products can vary depending on the specific dental materials used. Dental materials such as dental composites, dental cements, and dental adhesives may undergo degradation over time due to factors such as oral pH, mechanical stresses, and exposure to saliva and other oral fluids [38]. Here's an overview of some common dental materials and their potential degradation products:

3.1 Dental composites

Dental composites are tooth-colored restorative materials used to fill cavities and repair teeth. They typically consist of a resin matrix (monomers) and filler particles (such as glass or ceramic) [39]. Electrophoresis analysis of composite materials provides valuable insights into their composition, polymerization characteristics, and degradation products [2, 6, 7]. Electrophoresis techniques, such as PAGE or CE, are utilized to analyze the monomers present in composite materials. The analysis helps in assessing the efficiency of the polymerization process, the presence of unreacted monomers, and the identification of potential leachable components [40]. Additionally, electrophoresis can be used to investigate the degradation products of composite materials over time, providing information on their long-term stability and biocompatibility [41]. The charges and sizes of degradation products in dentistry are highly specific to the materials used. Electrophoresis can be used to analyze the degradation products, but it is not a common method in dental research or clinical practice. Dental materials are typically analyzed through other methods, such as spectroscopy, chromatography, scanning electron microscopy, and mechanical testing. While electrophoresis is a powerful tool for analyzing the degradation products of biomolecules such as DNA, RNA, and proteins, it is not commonly used for analyzing degradation products in dental materials. Dental materials have different chemical compositions, and their degradation products are typically assessed using other specialized analytical techniques [42]. Degradation of dental composites may result in:

- Leaching of monomers: Some monomers used in dental composites, such as bisphenol A glycidyl methacrylate (Bis-GMA), triethylene glycol dimethacrylate (TEGDMA), and urethane dimethacrylate (UDMA), may undergo leaching into the oral environment over time.
- Filler particle breakdown: The filler particles in dental composites may undergo wear or degradation, leading to the release of fine particles into the oral cavity.

These monomers are polymerized to form dental materials, which play a crucial role in their mechanical and adhesive properties [42]. It's important to note that dental materials and their degradation mechanisms are continuously being studied and improved to enhance their performance and longevity in dental restorations and treatments. Dental research aims to develop materials that withstand the oral environment and minimize potential adverse effects from degradation products.

3.2 Dental cements

Dental cements are materials used to bond various dental restorations to teeth, essential in various dental applications, including cementation of crowns, bridges, and orthodontic appliances [43]. They can be classified into different types, such as resin-based cements, glass ionomer cements, and zinc phosphate cements. Electrophoresis analysis aids in the evaluation of the composition, setting reaction, and mechanical properties of dental cements. Electrophoretic techniques, such as PAGE or CE, are employed to assess the composition of dental cements and identify the presence of various components, such as resin monomers, initiators, fillers, and additives. This analysis helps in understanding the role of each component in the cement's properties and performance. Furthermore, electrophoresis can be utilized to study the setting reaction of dental cements, allowing for the identification and quantification of reaction by-products and the assessment of the cement's final properties [44]. In addition, the degradation products may include [45]:

- Ion release: Glass ionomer cements, for example, can release fluoride and other ions over time, which can have beneficial effects on the adjacent tooth structure.

- Chemical reactions: Some cements may undergo chemical reactions leading to changes in their properties over time.

3.3 Impression materials

Impression materials are used to capture the precise dental structures for fabricating prosthetic devices and dental restorations [46]. Electrophoresis analysis of impression materials enables the evaluation of their composition, polymerization characteristics, and potential leakage [47]. Electrophoretic techniques, such as PAGE or CE, are employed to analyze the composition of impression materials, including the identification and quantification of various components, such as polymers, initiators, and accelerators. The analysis helps in assessing the material's consistency, stability, and performance. Additionally, electrophoresis can be used to study the polymerization process of impression materials, evaluating the efficiency and completeness of the polymerization reaction [48].

4. Electrophoresis analysis of oral disease-related proteins

4.1 Periodontal disease biomarkers

Periodontal diseases, including gingivitis and periodontitis, are characterized by inflammation and destruction of the periodontal tissues [49]. Biomarkers are specific molecules, proteins, or genes that can indicate the presence or severity of a disease. In the context of periodontal disease, biomarkers can help in diagnosis, prognosis, and monitoring of the progression of the disease. Protein patterns depict the collective arrangement of proteins within a biological sample, such as GCF or saliva [50]. In the study of periodontal diseases, techniques such as 2D gel electrophoresis and mass spectrometry are commonly employed to examine the protein patterns within GCF or saliva samples [51]. By comparing protein profiles from healthy individuals and those with periodontal disease, researchers can spot potential biomarkers linked to the condition. Electrophoresis analysis aids in the identification and characterization of biomarkers associated with periodontal diseases. This analysis allows for the identification of disease-specific proteins or protein patterns that can serve as potential biomarkers for periodontal diseases [51, 52]. Here are some example proteins and genes that have been studied as potential biomarkers for periodontal disease:

4.1.1 Example proteins as biomarkers

- Matrix metalloproteinases (MMPs): MMPs are enzymes that play a role in tissue remodeling and degradation of extracellular matrix components. Elevated levels of certain MMPs, such as MMP-8 and MMP-9, have been associated with periodontal tissue destruction [53].

- C-reactive protein (CRP): CRP is an acute-phase protein produced by the liver in response to inflammation. Increased levels of CRP have been linked to periodontal disease and may indicate systemic inflammation [54].

- Interleukins (ILs): Various interleukins, such as IL-1β, IL-6, IL-8, and IL-17, are involved in the immune response and have been implicated in periodontal inflammation [55].

- Tumor necrosis factor-alpha (TNF-α): TNF-α is a pro-inflammatory cytokine that plays a role in the inflammatory process of periodontal disease [56].

4.1.2 Example genes as biomarkers

- Interleukin-1 gene cluster (IL1): Genetic variations in the IL1 gene cluster have been associated with increased susceptibility to severe periodontitis [57].

- Toll-like receptor genes (TLRs): TLRs are involved in recognizing microbial components and initiating the immune response. Genetic variations in TLR genes have been linked to periodontal disease susceptibility [58].

- Human leukocyte antigen (HLA) Genes: Certain HLA genotypes have been associated with increased susceptibility to periodontal disease [59].

Quantitative analysis by electrophoresis is valuable for comparing protein expression levels between different groups of samples, such as healthy and diseased individuals, and for identifying potential biomarkers associated with specific conditions such as periodontal disease [60]. However, it's essential to ensure that the quantification process is accurate and reliable by using appropriate controls and validation methods. There are different methods for quantitative analysis in electrophoresis [1, 2, 60]:

1. Densitometry: In traditional gel electrophoresis, such as SDS-PAGE or 2DE, densitometry can be used to measure the intensity of protein bands on the gel. The densitometry data can be quantified using imaging software, providing relative information about protein abundance.

2. Western blotting (Immunoblotting): In Western blotting, after electrophoresis, the separated proteins are transferred to a membrane, and specific antibodies are used to detect and quantify the target proteins. The signal intensity of the protein bands on the membrane corresponds to their abundance in the sample.

3. Difference gel electrophoresis (DIGE): DIGE is a technique that allows for the simultaneous comparison of multiple samples in a single gel run. Different samples are labeled with different fluorescent dyes, and their protein patterns can be quantified and compared.

4. Capillary electrophoresis: In CE, samples are separated in a capillary tube, and the detector measures the migration times and peak heights of the analytes. This information can be used for quantitative analysis.

4.2 Dental caries biomarkers

Dental caries, commonly known as tooth decay, is a prevalent oral disease caused by the demineralization of tooth structures [61]. Electrophoresis analysis facilitates the identification and characterization of biomarkers associated with dental caries [62]. Electrophoretic techniques, such as PAGE or CE, are employed to analyze proteins present in dental plaque, saliva, or biofilms. This analysis helps in identifying specific proteins or protein profiles associated with cariogenic bacteria or host response to dental caries. By comparing the protein expression patterns between caries-free individuals and those with active caries, potential biomarkers for dental caries can be identified. Electrophoresis analysis of dental caries biomarkers contributes to the development of diagnostic tools for early caries detection, risk assessment, and personalized preventive strategies [63].

4.3 Oral cancer biomarkers

Oral cancer encompasses a range of malignancies affecting the oral cavity, including the lips, tongue, gingiva, and oral mucosa [64]. Electrophoresis analysis plays a crucial role in the identification and characterization of oral cancer biomarkers [65]. Using techniques such as PAGE or 2DE, researchers can separate and analyze proteins present in oral cancer tissues, saliva, or serum. This analysis aids in the identification of differentially expressed proteins that are associated with oral cancer development, progression, or metastasis. By comparing protein expression patterns between healthy individuals and those with oral cancer, potential biomarkers can be discovered. Electrophoresis analysis of oral cancer biomarkers contributes to the development of early detection methods, prognostic indicators, and targeted therapies for oral cancer management [66].

5. Advantages and limitations of electrophoresis in dentistry research

5.1 Advantages

5.1.1 High separation efficiency

Electrophoresis offers high separation efficiency, allowing for the resolution of complex mixtures of molecules based on their charge, size, or IP. This enables researchers to analyze and identify individual components within dental materials or disease-related proteins with precision [1].

5.1.2 Versatility

Electrophoresis techniques, such as PAGE, agarose gel electrophoresis, CE, and 2DE, provide a range of options for different types of analyses. This versatility allows researchers to adapt the technique to specific research objectives and sample requirements [1].

5.1.3 Quantification and comparison

Electrophoresis can be used for the quantification of proteins or other analytes present in dental materials or biological samples. By comparing the expression levels of specific proteins or the presence of certain components, researchers can assess differences between healthy and diseased states, evaluate the effectiveness of dental treatments, and monitor disease progression [2]. Quantifying dental materials and biological samples using electrophoresis typically involves measuring the intensity of bands or peaks on the electrophoresis gel or capillary. The quantification process can vary depending on the type of electrophoresis and the specific materials or biomolecules being analyzed [60]. Here are general steps for quantifying dental materials and biological samples using two common types of electrophoresis:

Quantification using SDS-PAGE for protein analysis [2, 4]:

1. Gel electrophoresis: Perform SDS-PAGE to separate the proteins in the sample. Use appropriate molecular weight markers to estimate the size of protein bands.

2. Gel staining: After electrophoresis, stain the gel with a protein-specific stain, such as Coomassie Brilliant Blue or a fluorescent dye. The stain will bind to the proteins, allowing them to be visualized.

3. Gel imaging: Capture an image of the stained gel using a gel documentation system or a high-resolution scanner.

4. Densitometry: Use image analysis software to quantify the intensity of protein bands. Select the region of interest (band) on the gel, and the software will calculate the integrated optical density or intensity of the band.

5. Standard curve: For more accurate quantification, create a standard curve using known concentrations of purified protein standards. The standard curve can be used to convert the measured intensity of protein bands into protein concentrations.

Quantification using agarose gel electrophoresis for nucleic acid analysis [1, 2]:

1. Agarose gel electrophoresis: Run the DNA or RNA samples on an agarose gel to separate the nucleic acids based on size.

2. Gel staining: After electrophoresis, stain the gel with a DNA-specific dye (e.g., ethidium bromide or SYBR Green) to visualize the DNA bands.

3. Gel imaging: Capture an image of the stained gel using a gel documentation system or a high-resolution scanner.

4. Densitometry (optional for DNA): For more accurate quantification of DNA bands, you can use densitometry software to measure the intensity of DNA bands. However, densitometry is not commonly used for DNA quantification; instead, DNA quantification is typically performed using spectrophotometry or fluorometry before electrophoresis.

Quantification of other biomolecules (e.g., RNA, proteins, or DNA in capillary electrophoresis) [1, 24, 25].

For some advanced techniques such as capillary electrophoresis, quantification can be performed using specialized software that analyzes the migration times and peak heights of the analytes. The software can compare the migration times and peak areas of the sample peaks with those of standard solutions of known concentrations. It's important to note that the accuracy of quantification using electrophoresis depends on several factors, such as the quality of the gel or capillary run, the staining and detection methods, and the selection of appropriate standards for calibration. Always validate the obtained quantification results and consider using multiple complementary methods for quantification to ensure accuracy and reliability.

5.1.4 Protein characterization

Electrophoresis facilitates the characterization of proteins by providing information on their molecular weight, IP, and charge heterogeneity. This information aids in the identification of disease-related proteins, protein modifications, and protein-protein interactions, contributing to a deeper understanding of the molecular mechanisms underlying oral diseases [20]. Identification of disease-related proteins, protein modifications, and protein-protein interactions using electrophoresis often involves a combination of electrophoretic techniques with other analytical methods [67, 68]. Here's an overview of how each of these aspects can be achieved.

5.1.4.1 Identification of disease-related proteins

Electrophoresis alone does not provide direct identification of disease-related proteins. Instead, it is often used as a first step to separate complex protein mixtures, followed by additional methods for protein identification [1, 67]. Two commonly used techniques for protein identification after electrophoresis are mass spectrometry and Western blotting.

- Mass spectrometry (MS): After gel electrophoresis, the proteins of interest are typically extracted from the gel, digested into peptides, and analyzed by MS. The latter can provide information about the masses and sequences of the peptides, allowing researchers to identify the corresponding proteins through database searches.

- Western blotting (Immunoblotting): Western blotting is used to detect and quantify specific proteins in a complex mixture. After gel electrophoresis, proteins are transferred to a membrane and probed with specific antibodies against the proteins of interest. The antibodies bind to their targets, and the presence of the protein is detected through chemiluminescence or fluorescence.

5.1.4.2 Identification of protein modifications

Electrophoresis can be combined with specific staining or immunodetection techniques to identify protein modifications, such as posttranslational modifications (PTMs) [60]. Some examples of PTM identification are:

- Phosphorylation: Phosphorylated proteins can be detected using phospho-specific antibodies in Western blotting or specific stains, such as Pro-Q Diamond, in 2D gel electrophoresis.

- Glycosylation: Glycosylated proteins can be identified using lectin staining in gel electrophoresis or through glycoprotein-specific antibodies.

5.1.4.3 Identification of protein-protein interactions

Electrophoresis itself is not a direct method for identifying protein-protein interactions. However, it can be used as a preliminary step in techniques that investigate protein-protein interactions [1, 25]. Two commonly used methods for this purpose are co-immunoprecipitation (Co-ip) and pull-down assays.

- Co-immunoprecipitation (Co-ip): In Co-ip, proteins are first cross-linked or stabilized to preserve protein-protein interactions. The target protein is immunoprecipitated using specific antibodies, and the co-immunoprecipitated proteins are then separated by gel electrophoresis and identified using MS or Western blotting.

- Pull-down assays: Pull-down assays use affinity chromatography to isolate one protein from a complex mixture, along with its interacting partners. After electrophoresis, the interacting proteins can be identified using mass spectrometry or Western blotting.

5.1.5 Biomarker discovery

Electrophoresis plays a crucial role in the discovery of biomarkers associated with dental diseases. By analyzing protein expression patterns or identifying disease-specific proteins, researchers can identify potential biomarkers for early detection, diagnosis, prognosis, and monitoring of oral diseases. This information has the potential to improve patient outcomes and guide personalized treatment approaches [21].

5.2 Limitations

5.2.1 Sample complexity

One of the main limitations of electrophoresis is the complexity of samples, particularly in biological fluids or tissues. The presence of a wide range of proteins, nucleic acids, and other biomolecules can hinder the separation and identification of specific components. Additional sample preparation steps, such as protein enrichment or fractionation, may be required to overcome this limitation [69].

5.2.2 Sensitivity

The sensitivity of electrophoresis techniques can vary depending on the type of analysis and detection method employed. Some low-abundance proteins or analytes may be challenging to detect, especially in complex samples. Sensitivity limitations can affect the ability to identify specific biomarkers or quantify analytes accurately [70].

5.2.3 Interpretation of results

Electrophoresis provides a visual representation of separated components, but the interpretation of results requires expertise and careful analysis. The identification of specific proteins or analytes often requires additional techniques, such as mass spectrometry or immunoassays, to confirm their identity. Additionally, the presence of posttranslational modifications or protein isoforms can complicate the interpretation of electrophoresis results [71].

5.2.4 Standardization and reproducibility

Achieving consistent and reproducible results in electrophoresis can be challenging due to variations in gel preparation, running conditions, staining methods, and data analysis. Standardization of protocols and rigorous quality control measures are necessary to ensure reliable and comparable results across different studies and laboratories [72].

5.2.5 Technical expertise and equipment requirements

Electrophoresis techniques require specialized equipment, such as gel electrophoresis apparatus, power supplies, and imaging systems. Additionally, the interpretation of results and data analysis often necessitate expertise in molecular biology, biochemistry, and biostatistics. These requirements may limit the accessibility of electrophoresis to certain research settings or require collaboration with experienced researchers or core facilities [73].

6. Discussion

The use of electrophoresis in dental research brings notable advantages and potential applications for analyzing dental materials and proteins linked to oral diseases. The prior sections of this chapter have presented how electrophoresis enhances our comprehension of dental materials, oral diseases, and personalized patient care in dentistry. Electrophoresis emerges as an effective technique for dental materials analysis. Employing diverse methods such as PAGE, agarose gel electrophoresis, and 2DE, researchers have effectively separated and characterized dental material components based on their charge, size, or IP [1, 2]. This insight has proven invaluable in understanding the composition, purity, and structural attributes of dental materials, offering an assessment of their quality, performance, and compatibility.

Furthermore, electrophoresis techniques have facilitated the identification of specific components within dental materials, such as resin composites, cements, and impression materials, contributing to the development of improved dental materials with enhanced properties and clinical outcomes [63]. Apart from dental materials, electrophoresis has played a significant role in investigating proteins connected to oral diseases [65]. By examining protein expression patterns and identifying disease-specific proteins, researchers have been able to discover and validate potential biomarkers related to oral issues such as periodontal disease, dental caries, and oral cancer. Techniques such as 2DE and IEF have enabled the dissection and characterization of intricate protein mixtures, furnishing crucial insights into protein modifications, interactions, and deviations during disease states. This knowledge has the potential to transform clinical practice by aiding early detection, diagnosis, monitoring of oral problems, and guiding tailored treatment plans for individual patients [69].

While electrophoresis has demonstrated great efficacy in dentistry research, it's essential to recognize its limitations. The intricacy of samples, especially in biological fluids or tissues, poses challenges in segregating and identifying specific components using electrophoresis techniques [69]. Overcoming this challenge might require preparatory steps such as protein enrichment or fractionation. Moreover, the sensitivity of electrophoresis techniques can fluctuate, making the detection of low-abundance proteins or analytes challenging, particularly in complex samples. Therefore, supplementary techniques such as MS or immunoassays might be necessary to validate and confirm the identity of specific proteins or biomarkers [62, 71].

To propel the field forward, upcoming research should address these limitations and explore novel applications of electrophoresis in dentistry [74]. Integrating multi-omics approaches, encompassing genomics, transcriptomics, proteomics, and metabolomics, can furnish a more comprehensive grasp of the molecular mechanisms behind dental issues and the influence of dental materials [75]. Validating identified biomarkers across larger patient groups and diverse populations is pivotal for their practical use and integration [76]. Moreover, incorporating artificial intelligence (AI) and machine learning algorithms can enhance data analysis, pattern recognition, and predictive modeling, leading to more precise and individualized diagnostic and treatment strategies in the realm of dentistry [77, 78].

7. Future perspectives and conclusions

7.1 Future perspectives

Electrophoresis has been a valuable tool in dentistry research, enabling the analysis of dental materials and oral disease-related proteins. Looking ahead, several future perspectives can be envisioned for the application of electrophoresis in dentistry.

7.1.1 Advancements in electrophoresis techniques

Electrophoresis techniques are continually evolving, driven by technological advancements. Future improvements may include the development of novel gel matrices, better detection methods, and enhanced automation. These advancements would lead to increased sensitivity, resolution, and throughput, enabling more comprehensive and efficient analysis of dental materials and oral disease-related proteins [74].

7.1.2 Integration of multi-omics approaches

The integration of different omics approaches, such as genomics, transcriptomics, proteomics, and metabolomics, holds great potential for advancing dentistry research. By combining electrophoresis analysis with other high-throughput technologies, researchers can obtain a more comprehensive understanding of the molecular mechanisms underlying dental diseases and the effects of dental materials [75].

7.1.3 Biomarker validation and clinical translation

The discovery of potential biomarkers through electrophoresis analysis requires further validation and translation into clinical practice. Future studies should focus on validating the identified biomarkers in larger patient cohorts and diverse populations. Moreover, efforts should be made to develop reliable and user-friendly diagnostic assays that can be readily implemented in dental clinics [76].

7.1.4 Integration of artificial intelligence and machine learning

The integration of AI and machine learning algorithms can enhance data analysis, pattern recognition, and predictive modeling in dentistry research. AI-driven approaches can aid in the identification of complex protein patterns, the prediction of disease progression, and the optimization of dental material properties [77, 78].

8. Conclusions

Electrophoresis has emerged as a valuable asset in dentistry research, shedding light on dental materials and proteins linked to oral diseases. Its benefits encompass efficient separation, versatility, quantification capabilities, protein understanding, and biomarker discovery. Nonetheless, it's essential to acknowledge limitations such as complex samples, sensitivity, result interpretation, standardization, and technical expertise. Looking ahead, progress in electrophoresis techniques, incorporating multi-omics methods, validating biomarkers, clinical adaptation, and utilizing AI and machine learning offer exciting avenues to amplify electrophoresis' role in dentistry research. By embracing these forthcoming possibilities and tackling current challenges, researchers can further leverage electrophoresis as a potent tool for studying dental materials and oral disease-related proteins. This, in turn, will advance diagnostics, treatments, and patient care in the field of dentistry.

Conflict of interest

The authors declare no conflict of interest.

Author details

Aida Meto[1,2]* and Agron Meto[1]

1 Faculty of Dental Sciences, Department of Dentistry, University of Aldent, Tirana, Albania

2 Clinical Microbiology, School of Dentistry, University of Modena and Reggio Emilia, Modena, Italy

*Address all correspondence to: aida.meto@ual.edu.al

References

[1] Sonagra AD, Dholariya SJ. Electrophoresis. In: StatPearls [Internet]. Treasure Island, FL: StatPearls Publishing; Jan–31 Jul 2023. PMID: 36251838

[2] Nagpal M, Sood S. Role of electrophoresis in dental research. Indian Journal of Oral Sciences. 2013;**4**(1):9-13. DOI: 10.4103/0976-6944.113224

[3] Pawlowska E, Poplawski T, Ksiazek D, Szczepanska J, Blasiak J. Genotoxicity and cytotoxicity of 2-hydroxyethyl methacrylate. Mutation Research. 2010;**696**(2):122-129. DOI: 10.1016/j.mrgentox.2009.12.019

[4] Green MR, Sambrook J. Polyacrylamide gel electrophoresis. Cold Spring Harbor Protocols. 2020;**2020**(12):525-532. DOI: 10.1101/pdb.prot100412. PMID: 33262236

[5] Ramautar R. Capillary electrophoresis-mass spectrometry for clinical metabolomics. Advances in Clinical Chemistry. 2016;**74**:1-34. DOI: 10.1016/bs.acc.2015.12.002

[6] Meleady P. Two-dimensional gel electrophoresis and 2D-DIGE. Methods in Molecular Biology. 2023;**2596**:3-15. DOI: 10.1007/978-1-0716-2831-7_1

[7] Wang S, Guo L, Seneviratne CJ, et al. Biofilm formation of salivary microbiota on dental restorative materials analyzed by denaturing gradient gel electrophoresis and sequencing. Dental Materials Journal. 2014;**33**(3):325-331. DOI: 10.4012/dmj.2013-152

[8] Vervliet P, de Nys S, Boonen I, et al. Qualitative analysis of dental material ingredients, composite resins and sealants using liquid chromatography coupled to quadrupole time of flight mass spectrometry. Journal of Chromatography. A. 2018;**1576**:90-100. DOI: 10.1016/j.chroma.2018.09.039

[9] Naik VB, Jain AK, Rao RD, Naik BD. Comparative evaluation of clinical performance of ceramic and resin inlays, onlays, and overlays: A systematic review and meta-analysis. Journal of Conservative Dentistry. 2022;**25**(4):347-355. DOI: 10.4103/jcd.jcd_184_22

[10] Geraldeli S, Maia Carvalho LA, de Souza Araújo IJ, et al. Incorporation of arginine to commercial orthodontic light-cured resin cements-physical, adhesive, and antibacterial properties. Materials (Basel). 2021;**14**(16):4391. DOI: 10.3390/ma14164391

[11] Bacchi A, Spazzin AO, de Oliveira GR, Pfeifer C, Cesar PF. Resin cements formulated with thio-urethanes can strengthen porcelain and increase bond strength to ceramics. Journal of Dentistry. 2018;**73**:50-56. DOI: 10.1016/j.jdent.2018.04.002

[12] Nakonieczny DS, Kern F, Dufner L, Antonowicz M, Matus K. Alumina and zirconia-reinforced polyamide PA-12 composites for biomedical additive manufacturing. Materials (Basel). 2021;**14**(20):6201. DOI: 10.3390/ma14206201

[13] Görg A, Weiss W, Dunn MJ. Current two-dimensional electrophoresis technology for proteomics. Proteomics. 2005;**5**(3):826-827. DOI: 10.1002/pmic.200401031

[14] Mangum JE, Kon JC, Hubbard MJ. Proteomic analysis of dental tissue microsamples. Methods in Molecular

Biology. 2010;**666**:309-325. DOI: 10.1007/978-1-60761-820-1_19

[15] Beeley JA. Clinical applications of electrophoresis of human salivary proteins. Journal of Chromatography. 1991;**569**(1-2):261-280. DOI: 10.1016/0378-4347(91)80233-3

[16] Jiang Q, Liu J, Chen L, Gan N, Yang D. The Oral microbiome in the elderly with dental caries and health. Frontiers in Cellular and Infection Microbiology. 2019;**8**:442. DOI: 10.3389/fcimb.2018.00442

[17] Komuro T, Nakamura M, Tsutsumi H, Mukoyama R. Gender determination from dental pulp by using capillary gel electrophoresis of amelogenin locus. The Journal of Forensic Odonto-Stomatology. 1998;**16**(2):23-26

[18] Zhang XS, Proctor GB, Garrett JR, Schulte BA, Shori DK. Use of lectin probes on tissues and sympathetic saliva to study the glycoproteins secreted by rat submandibular glands. The Journal of Histochemistry and Cytochemistry. 1994;**42**(9):1261-1269. DOI: 10.1177/42.9.8064133

[19] Kinoshita E, Kinoshita-Kikuta E, Koike T. Separation and detection of large phosphoproteins using Phos-tag SDS-PAGE. Nature Protocols. 2009;**4**(10):1513-1521. DOI: 10.1038/nprot.2009.154

[20] Sánchez-Medrano AG, Martinez-Martinez RE, Soria-Guerra R, Portales-Perez D, Bach H, Martinez-Gutierrez F. A systematic review of the protein composition of whole saliva in subjects with healthy periodontium compared with chronic periodontitis. PLoS One. 2023;**18**(5):e0286079. DOI: 10.1371/journal.pone.0286079

[21] Khurshid Z, Zafar MS, Khan RS, Najeeb S, Slowey PD, Rehman IU. Role of salivary biomarkers in Oral cancer detection. Advances in Clinical Chemistry. 2018;**86**:23-70. DOI: 10.1016/bs.acc.2018.05.002

[22] Tzimas K, Pappa E. Saliva Metabolomic profile in dental medicine research: A narrative review. Metabolites. 2023;**13**(3):379. DOI: 10.3390/metabo13030379

[23] Gardner A, Carpenter G, So PW. Salivary metabolomics: From diagnostic biomarker discovery to investigating biological function. Metabolites. 2020;**10**(2):47. DOI: 10.3390/metabo10020047

[24] Green MR, Sambrook J. Analysis of DNA by agarose gel electrophoresis. Cold Spring Harbor Protocols. 2019;**2019**(1):6-15. DOI: 10.1101/pdb.top100388

[25] Kurien BT, Scofield RH. Extraction of proteins from gels: A brief review. Methods in Molecular Biology. 2012;**869**:403-405. DOI: 10.1007/978-1-61779-821-4_33

[26] Slater GW. DNA gel electrophoresis: The reptation model(s). Electrophoresis. 2009;**30**(Suppl 1):S181-S187. DOI: 10.1002/elps.200900154

[27] Chhattani B, Kulkarni P, Agrawal N, Mali S, Kumar A, Thakur NS. Comparative evaluation of antimicrobial efficacy of silver diamine fluoride, chlorhexidine varnish with conventional fluoride varnish as a caries arresting agent. An in vivo sodium dodecyl sulfate-polyacrylamide gel electrophoresis study. Journal of the Indian Society of Pedodontics and Preventive Dentistry. 2021;**39**(4):398-402. DOI: 10.4103/jisppd.jisppd_246_21

[28] Ivessa AS. Directionality of replication fork movement determined by two-dimensional native-native DNA agarose gel electrophoresis. Methods in Molecular Biology. 2013;**1054**:83-103. DOI: 10.1007/978-1-62703-565-1_5

[29] Gabriel M, Zentner A. Sodium dodecyl sulfate agarose gel electropheresis and electroelution of high molecular weight human salivary mucin. Clinical Oral Investigations. 2005;**9**(4):284-286. DOI: 10.1007/s00784-005-0007-2

[30] Elbashir AA, Elgorashe REE, Alnajjar AO, Aboul-Enein HY. Application of capillary electrophoresis with Capacitively coupled contactless conductivity detection (CE-C4D): 2017-2020. Critical Reviews in Analytical Chemistry. 2022;**52**(3):535-543. DOI: 10.1080/10408347.2020.1809340

[31] Ishikawa S, Sugimoto M, Kitabatake K, et al. Identification of salivary metabolomic biomarkers for oral cancer screening. Scientific Reports. 2016;**6**:31520. DOI: 10.1038/srep31520

[32] Ishikawa S, Sugimoto M, Edamatsu K, Sugano A, Kitabatake K, Iino M. Discrimination of oral squamous cell carcinoma from oral lichen planus by salivary metabolomics. Oral Diseases. 2020;**26**(1):35-42. DOI: 10.1111/odi.13209

[33] Arrer E, Patsch W, Meco C, Oberascher G, Albegger K. Two-dimensional electrophoresis (2-DE). Head & Neck. 2005;**27**(4):344-345. DOI: 10.1002/hed.20185

[34] Muroi M, Osada H. Two-dimensional electrophoresis-cellular thermal shift assay (2DE-CETSA) for target identification of bioactive compounds. Methods in Enzymology. 2022;**675**:425-437. DOI: 10.1016/bs.mie.2022.07.018

[35] Dowling P, O'Sullivan EM. Analysis of the saliva proteome using 2D-DIGE. Methods in Molecular Biology. 2023;**2596**:169-174. DOI: 10.1007/978-1-0716-2831-7_13

[36] Yu S, Yan C, Hu X, He B, Jiang Y, He Q. Isoelectric focusing on microfluidic paper-based chips. Analytical and Bioanalytical Chemistry. 2019;**411**(21):5415-5422. DOI: 10.1007/s00216-019-02008-5

[37] Widbiller M, Schweikl H, Bruckmann A, Rosendahl A, Hochmuth E, Lindner SR, et al. Shotgun proteomics of human dentin with different Prefractionation methods. Scientific Reports. 2019;**9**(1):4457. DOI: 10.1038/s41598-019-41144-x

[38] Santerre JP, Shajii L, Leung BW. Relation of dental composite formulations to their degradation and the release of hydrolyzed polymeric-resin-derived products. Critical Reviews in Oral Biology and Medicine. 2001;**12**(2):136-151. DOI: 10.1177/10454411010120020401

[39] Ilie N, Hickel R. Resin composite restorative materials. Australian Dental Journal. 2011;**56**(Suppl 1):59-66. DOI: 10.1111/j.1834-7819.2010.01296.x

[40] Tadin A, Galic N, Mladinic M, Marovic D, Kovacic I, Zeljezic D. Genotoxicity in gingival cells of patients undergoing tooth restoration with two different dental composite materials. Clinical Oral Investigations. 2014;**18**(1):87-96. DOI: 10.1007/s00784-013-0933-3

[41] Zhu XD, Fan HS, Zhao CY, et al. Competitive adsorption of bovine serum albumin and lysozyme on characterized calcium phosphates by polyacrylamide gel electrophoresis method. Journal of

Materials Science. Materials in Medicine. 2007;**18**(11):2243-2249. DOI: 10.1007/s10856-006-0057-2

[42] Vervliet P, De Nys S, Duca RC, et al. Identification of chemicals leaching from dental resin-based materials after in vitro chemical and salivary degradation. Dental Materials. 2022;**38**(1):19-32. DOI: 10.1016/j.dental.2021.10.006

[43] Swift EJ Jr. Contemporary dental cements. Journal of Esthetic and Restorative Dentistry. 2012;**24**(6): 365-366. DOI: 10.1111/jerd.12003

[44] Yamamura T, Kitagawa S, Ohtani H. Separation of linear synthetic polymers in non-aqueous capillary zone electrophoresis using cationic surfactant. Journal of Chromatography. A. 2015;**1393**:122-127. DOI: 10.1016/j.chroma.2015.03.036

[45] Oilo G. Biodegradation of dental composites/glass-ionomer cements. Advances in Dental Research. 1992;**6**:50-54. DOI: 10.1177/08959374920060011701

[46] Punj A, Bompolaki D, Garaicoa J. Dental impression materials and techniques. Dental Clinics of North America. 2017;**61**(4):779-796. DOI: 10.1016/j.cden.2017.06.004

[47] Kim M, Siegler K, Tamariz J, et al. Identification and long-term stability of DNA captured on a dental impression wafer. Pediatric Dentistry. 2012;**34**(5):373-377

[48] Barker CS, Soro V, Dymock D, Sandy JR, Ireland AJ. Microbial contamination of laboratory constructed removable orthodontic appliances. Clinical Oral Investigations. 2014;**18**(9):2193-2202. DOI: 10.1007/s00784-014-1203-8

[49] Mariotti A, Hefti AF. Defining periodontal health. BMC Oral Health. 2015;**15**(Suppl 1):S6. DOI: 10.1186/1472-6831-15-S1-S6

[50] Barros SP, Williams R, Offenbacher S, Morelli T. Gingival crevicular fluid as a source of biomarkers for periodontitis. Periodontology 2000. 2000;**70**(1):53-64. DOI: 10.1111/prd.12107

[51] Li Z, Chen S, Liu C, Zhang D, Dou X, Yamaguchi Y. Quantification of periodontal pathogens cell counts by capillary electrophoresis. Journal of Chromatography. A. 2014;**1361**:286-290. DOI: 10.1016/j.chroma.2014.07.100

[52] Binti Badlishah Sham NI, Lewin SD, Grant MM. Proteomic investigations of In vitro and In vivo models of periodontal disease. Proteomics. Clinical Applications. 2020;**14**(3):e1900043. DOI: 10.1002/prca.201900043

[53] Sorsa T, Tjäderhane L, Konttinen YT, et al. Matrix metalloproteinases: Contribution to pathogenesis, diagnosis and treatment of periodontal inflammation. Annals of Medicine. 2006;**38**(5):306-321. DOI: 10.1080/07853890600800103

[54] Barbi W, Kumar S, Sinha S, Askari M, Priya S, Kumar SJ. Reliability of C-reactive protein as a biomarker for cardiovascular and Oral diseases in young and old subjects. Journal of Pharmacy & Bioallied Sciences. 2021;**13**(Suppl 2):S1458-S1461. DOI: 10.4103/jpbs.jpbs_251_21

[55] Cheng WC, Hughes FJ, Taams LS. The presence, function and regulation of IL-17 and Th17 cells in periodontitis. Journal of Clinical Periodontology. 2014;**41**(6):541-549. DOI: 10.1111/jcpe.12238

[56] Romero-Castro NS, Vázquez-Villamar M, Muñoz-Valle JF, et al. Relationship between TNF-α, MMP-8, and MMP-9 levels in gingival crevicular fluid and the subgingival microbiota in periodontal disease. Odontology. 2020;**108**(1):25-33. DOI: 10.1007/s10266-019-00435-5

[57] Borilova Linhartova P, Danek Z, Deissova T, et al. Interleukin gene variability and periodontal bacteria in patients with generalized aggressive form of periodontitis. International Journal of Molecular Sciences. 2020;**21**(13):4728. DOI: 10.3390/ijms21134728

[58] Gu Y, Han X. Toll-like receptor signaling and immune regulatory lymphocytes in periodontal disease. International Journal of Molecular Sciences. 2020;**21**(9):3329. DOI: 10.3390/ijms21093329

[59] Naito T, Cunha-Cruz J. Human leukocyte antigen combinations may be risk factors for periodontal disease. The Journal of Evidence-Based Dental Practice. 2005;**5**(1):43-44. DOI: 10.1016/j.jebdp.2005.01.014

[60] Salih E. Qualitative and quantitative proteome analysis of Oral fluids in health and periodontal disease by mass spectrometry. Methods in Molecular Biology. 2017;**1537**:37-60. DOI: 10.1007/978-1-4939-6685-1_3

[61] Selwitz RH, Ismail AI, Pitts NB. Dental caries. Lancet. 2007;**369**(9555):51-59. DOI: 10.1016/S0140-6736(07)60031-2

[62] Ahmad P, Hussain A, Siqueira WL. Mass spectrometry-based proteomic approaches for salivary protein biomarkers discovery and dental caries diagnosis: A critical review. Mass Spectrometry Reviews. 29 Nov 2022:e21822. DOI: 10.1002/mas.21822. PMID: 36444686 [Epub ahead of print]

[63] Wittig I, Braun HP, Schägger H. Blue native PAGE. Nature Protocols. 2006;**1**(1):418-428. DOI: 10.1038/nprot.2006.62

[64] Wong T, Wiesenfeld D. Oral Cancer. Australian Dental Journal. 2018;**63**(Suppl 1):S91-S99. DOI: 10.1111/adj.12594

[65] Kaur J, Jacobs R, Huang Y, Salvo N, Politis C. Salivary biomarkers for oral cancer and pre-cancer screening: A review. Clinical Oral Investigations. 2018;**22**(2):633-640. DOI: 10.1007/s00784-018-2337-x

[66] Yang Y, Huang J, Rabii B, Rabii R, Hu S. Quantitative proteomic analysis of serum proteins from oral cancer patients: Comparison of two analytical methods. International Journal of Molecular Sciences. 2014;**15**(8):14386-14395. DOI: 10.3390/ijms150814386

[67] Pelech S. Tracking cell signaling protein expression and phosphorylation by innovative proteomic solutions. Current Pharmaceutical Biotechnology. 2004;**5**(1):69-77. DOI: 10.2174/1389201043489666

[68] Kido J, Bando M, Hiroshima Y, et al. Analysis of proteins in human gingival crevicular fluid by mass spectrometry. Journal of Periodontal Research. 2012;**47**(4):488-499. DOI: 10.1111/j.1600-0765.2011.01458.x

[69] Hert DG, Fredlake CP, Barron AE. Advantages and limitations of next-generation sequencing technologies: A comparison of electrophoresis and non-electrophoresis methods. Electrophoresis. 2008;**29**(23):4618-4626. DOI: 10.1002/elps.200800456

[70] Breadmore MC, Wuethrich A, Li F, et al. Recent advances in enhancing the sensitivity of electrophoresis and electrochromatography in capillaries and microchips (2014-2016). Electrophoresis. 2017;**38**(1):33-59. DOI: 10.1002/elps.201600331

[71] Weiß F, van den Berg BH, Planatscher H, Pynn CJ, Joos TO, Poetz O. Catch and measure-mass spectrometry-based immunoassays in biomarker research. Biochimica et Biophysica Acta. 2014;**1844**(5):927-932. DOI: 10.1016/j.bbapap.2013.09.010

[72] Seifert H, Dolzani L, Bressan R, et al. Standardization and interlaboratory reproducibility assessment of pulsed-field gel electrophoresis-generated fingerprints of Acinetobacter baumannii. Journal of Clinical Microbiology. 2005;**43**(9):4328-4335. DOI: 10.1128/JCM.43.9.4328-4335.2005

[73] Csako G. Immunoelectrophoresis: A method with many faces. Methods in Molecular Biology. 2019;**1855**:249-268. DOI: 10.1007/978-1-4939-8793-1_21

[74] Holland LA, Casto-Boggess LD. Gels in microscale electrophoresis. The Annual Review of Analytical Chemistry (Palo Alto Calif). 2023;**16**(1):161-179. DOI: 10.1146/annurev-anchem-091522-080207

[75] Athieniti E, Spyrou GM. A guide to multi-omics data collection and integration for translational medicine. Computational and Structural Biotechnology Journal. 2022;**21**:134-149. DOI: 10.1016/j.csbj.2022.11.050

[76] Gürsoy UK, Kantarci A. Molecular biomarker research in periodontology: A roadmap for translation of science to clinical assay validation. Journal of Clinical Periodontology. 2022;**49**(6):556-561. DOI: 10.1111/jcpe.13617

[77] Carrillo-Perez F, Pecho OE, Morales JC, et al. Applications of artificial intelligence in dentistry: A comprehensive review. Journal of Esthetic and Restorative Dentistry. 2022;**34**(1):259-280. DOI: 10.1111/jerd.12844

[78] Alabi RO, Youssef O, Pirinen M, et al. Machine learning in oral squamous cell carcinoma: Current status, clinical concerns and prospects for future-a systematic review. Artificial Intelligence in Medicine. 2021;**115**:102060. DOI: 10.1016/j.artmed.2021.102060

Chapter 4

Molecular Techniques for Analysis of Biodiversity by Agarose Gel Electrophoresis

Estefanía García-Luque, Ana del Pino-Pérez and Enrique Viguera

Abstract

Molecular techniques based on DNA analysis have become an indispensable tool for the identification and classification of organisms, addressing the limitations of taxonomy based on morphological characters. There are different methods for the analysis of the variability of DNA which can provide unique genetic signatures capable of distinguishing closely related species, hybrid specimens or even individuals within the same species. Here we describe two methods that allow species identification by agarose gel electrophoresis separation techniques. DNA barcoding is a method of identifying any species based on a short DNA sequence amplified by PCR from a specific region of the genome, as most species have distinct genetic markers, or "barcodes", that are unique to them. By performing a bioinformatic analysis of the PCR-amplified barcode of an unknown sample against a database of known barcodes, it is possible to identify the species to which the sample belongs. On the other hand, Random Amplified Polymorphic DNA (RAPD) is used to detect genetic variation within a species. It is a PCR-based method that employs short, random primers to amplify DNA fragments from genomic DNA. The amplified fragments are then separated by gel electrophoresis and visualized as a banding pattern on the gel.

Keywords: DNA barcode, cytochrome oxidase, COI, RAPD, taxonomy, genetic diversity, STR

1. Introduction

DNA barcoding and RAPD (Random Amplified Polymorphic DNA) are two molecular biology techniques based on PCR used to analyze genetic variability within populations and species. These methods rely on the presence of genetic variation, which originates from mutation and recombination processes together with genetic drift and gene flow. Examining genetic variation within and between species provides insight into how organisms are related to each other and how they have evolved over time. Genetic variation can help determine the degree of relatedness between different species, identify common ancestors and reveal patterns of diversification and adaptation. It also helps to understand the genetic distinctiveness and uniqueness of different species, thus contributing to our understanding of biodiversity and the processes that shape it. We will discuss the application of these molecular markers along with others that can be used as alternatives in genetic studies. Among the most common options are those based on enzymatic digestion, such as RFLP (Restriction fragment length polymorphism), based on PCR, such as ISSR (Inter Simple Sequence Repeat), or on the combination of both procedures, such as AFLP (Amplified fragment length polymorphism). In this work we show the usefulness of using the DNA barcoding technique to identify food fraud in frozen fish samples. In addition, we use the RAPD technique for the identification of different Quercus species as an alternative technique in case there is not enough genetic variability to use the DNA barcoding technique.

2. DNA barcoding and uses

DNA barcoding is a molecular technique consisting of the amplification of short DNA sequences, between 400 and 800 bp, which allows the taxonomic identification of the genetic material under study [1, 2]. DNA barcoding is based on the analysis of a tissue sample for short DNA sequences of specific genes that act as genetic markers. Although there may be minor variations in highly conserved regions of DNA during evolution, the main idea of DNA barcoding is to identify a standardized region or "barcode" that shows enough variation between species to distinguish them. By comparing the DNA barcode sequences of unknown samples to a reference database, species can be identified and classified, even when traditional morphological identification is challenging or not feasible because of the sample scarcity or damaged samples. DNA barcoding is the preferred technique due to its cost-effectiveness and time requirements.

The great versatility of the barcoding technique makes it a very useful tool in different fields such as conservation biology, ecology, evolutionary biology and forensic analysis, among others. It has multiple applications such as species identification from any of the life stages of an organism (eggs, larvae, mature individuals or seeds), from damaged specimens, fecal or gut samples, to identify cryptic species (species that look similar but are genetically different), to promptly identify invasive species, to trace the biogeography of the species under study, studies on plant ecology, to control fraudulent trafficking of species or agricultural pests [3, 4]. It is also used by the pharmaceutical industry as a quality control or to check the correct labelling of foods in order to combat food fraud. **Figure 1** shows an analysis of DNA barcoding of fish species from several samples of small frozen fish fragments from a commercial supplier. DNA extraction, PCR amplification, barcode sequencing and comparison with molecular databases were performed to match the resulting species with the one marked on the sample labelling.

The use of short DNA sequences for species identification was first proposed by Hebert et al. [1] in 2003 at the University of Guelph, Canada. The mitochondrial cytochrome oxidase I (COI) gene was proposed as a candidate for establishing a global system for animal identification. The primer pair LCO1490 and HCO2198 (see **Table 1**) amplified a 658 bp region of this gene, sufficient for the identification of the specimens under study.

Subsequently, numerous studies were carried out to search for new genes for the use of the technique in other taxonomic groups such as plants [4, 7, 8] or fungi [9]. When studying plant DNA barcodes, three chloroplastidic genetic regions (rbcL, matK and trnH-psbA) and one nuclear region such as the Internal Transcribed Spacer (ITS) were selected as the standard barcode of choice for most plant and fungal [4, 7, 9]. Molecular markers used as barcodes in the identification of different taxonomic groups have been established by the Barcode of Life (CBOLD) or iBOLD

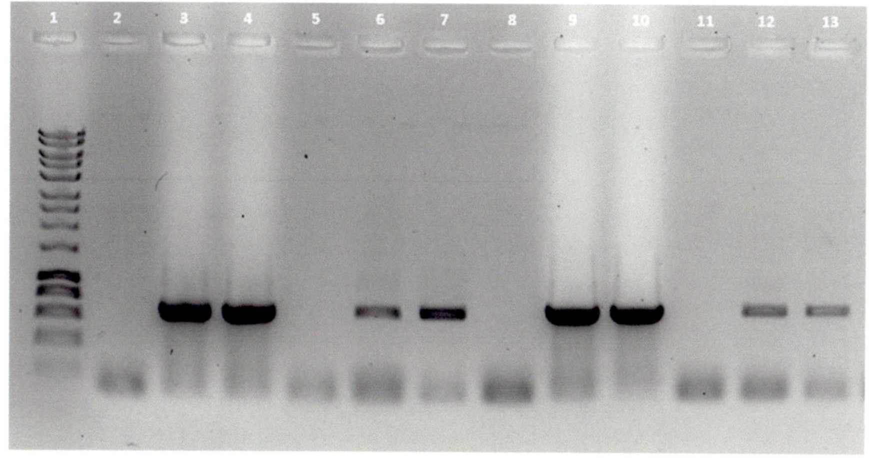

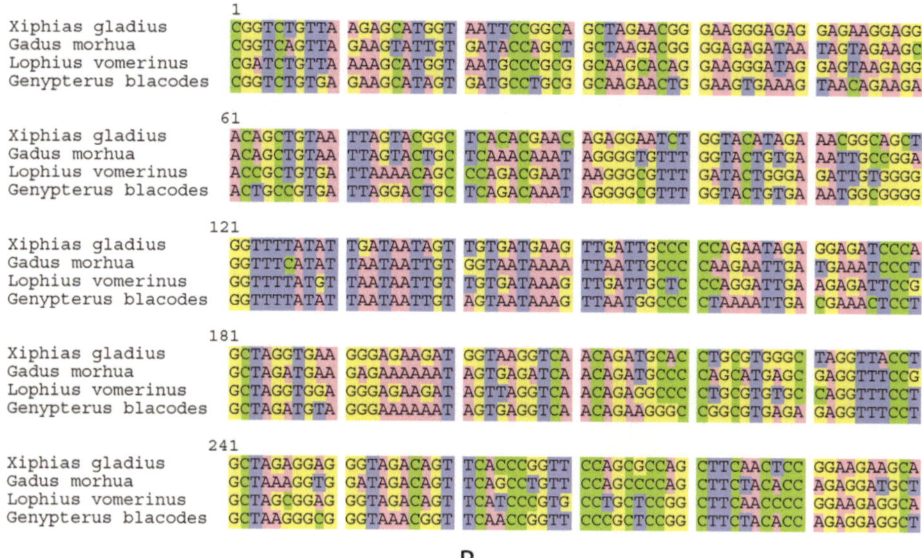

Figure 1.
Use of DNA barcoding to identify food fraud. (A) Agarose gel electrophoresis of COI amplification products from different commercial samples of frozen fish analyzed with primers FISHCOILBC_ ts and FISHCOIHBC_ts (See **Table 1***). Lane 1: molecular marker. Lanes 3, 4: Xiphias gladius; Lanes 6, 7: Gadus morhua; Lanes 9, 10: Lophius vomerinus; Lanes 12, 13: Genypterus blacodes. Lanes 2, 5, 8, 11: respective negative control samples. (B) Multiple sequence alignment of the barcodes used in this study by Seaview software [5].*

Consortium. The main steps of DNA barcoding are: (1) Sample collection, (2) Tissue sampling, (3) DNA isolation, (4) PCR amplification, (5) Sequencing and (6) analysis of the sequence obtained.

2.1 DNA barcoding technique

The DNA barcoding technique is based on the amplification of a DNA fragment by PCR with a primer pair complementary to the genomic region used as barcode. The amplification product is visualized by agarose gel electrophoresis. If a single

Primers name	Amplified gene	Sequence (5′-3′)	Amplified fragment size (pb)	PCR program
LCO1490	Mitochondrial gene. 5′ region of Subunit I of Cytochrome C oxidase (COI-5P) (invertebrate animals)	GGTCAACAAATCATAAAGATATTGG	≈700	(1)
HCO2198		TAAACTTCAGGGTGACCAAAAAATCA		
vF1i_t1	Mitochondrial gene. 5′ region of Subunit I of Cytochrome C oxidase (COI-5P) (non-fish vertebrate animals)	TCTCAACCAACCACAAAGACATTGG	≈700	(2)
vR1d_t1		TAGACTTCTGGGTGGCCRAARAAYCA		
FishCOILBC_ts	Mitochondrial gene. 5′ region of Cytochrome C oxidase subunit I (COI-5P) (fish vertebrate animals)	CACGACGTTGTAAAACGACTCAACYAA TCAYAAAGATATYGGCAC	≈700	(2)
FishCOIHBC_ts		GGATAACAATTTCACACAGGACTTCYGGGTGRCCRAARAATCA		
rbcLa f	Chloroplast gene. Ribulose 1,5-bisphosphate carboxylase/ oxygenase (rbcL) (plants)	ATGTCACCACAAACAGAGACTAAAGC	≈700	(2)
rbcLa rev		GTAAAATCAAGTCCACCRCG		
matk-3F	Chloroplast gene. Maturase K (matK) (angiosperms)	CGTACAGTACTTTTGTGTTTACGAG	≈800	(3)
matk-1R		ACCCAGTCCATCTGGAAATCTTGGTTC		
Gym_R1B_Fwd	Chloroplast gene. Maturase K (matK) (angiosperms)	TCATCCRGAAATTTTGGTKCG	≈800	(4)
Gym_F1B_Rev		ATMGTACTTTTATGTTTACARGC		
Eph_F	Chloroplast gene. Maturase K (matK) (Ephedraceae family)	TCATTCAGAGCTGTTAGTTAG	≈800	(5)
Eph_R		ATCGTACTTTTATGCTTACAGGC		
nrITS2-S2F	Nuclear gene. Internal transcription spacer (ITS) (plants)	ATGCGATACTTGGTGTGAAT	≈400	(5)
nrITS2-S3 Rev		GACGCTTCTCCAGACTACAAT		

DOI: http://dx.doi.org/10.5772/intechopen.1002268

Primers name	Amplified gene	Sequence (5′-3′)	Amplified fragment size (pb)	PCR program
TufA Fwd	Chloroplast gene. Elongation factor EF-Tu (Seaweed)	TGAAACAGAAMAWCGTCATTATGC	—	(1)
TufA Rev		CCTTCNCGAATMGCRAAWCGC		
ITS1 F	Nuclear gene. Internal Transcription Spacer (ITS) (Fungi)	CCGTAGGTGAACCTGCGG	—	(5)
ITS4 R		CCTCCGCTTATTGATATGC		
ITS1F_(Gad)	Nuclear gene. Internal transcription spacer (ITS) (Lichens)	TTGGTCATTTAGAGGAAGTA	—	(5)
ITS4 R		CCTCCGCTTATTGATATGC-		

"-": not determined in our study.

Table 1.
Pair of primers selected for different taxonomic groups. Data described are obtained from reference [6] and our own results.

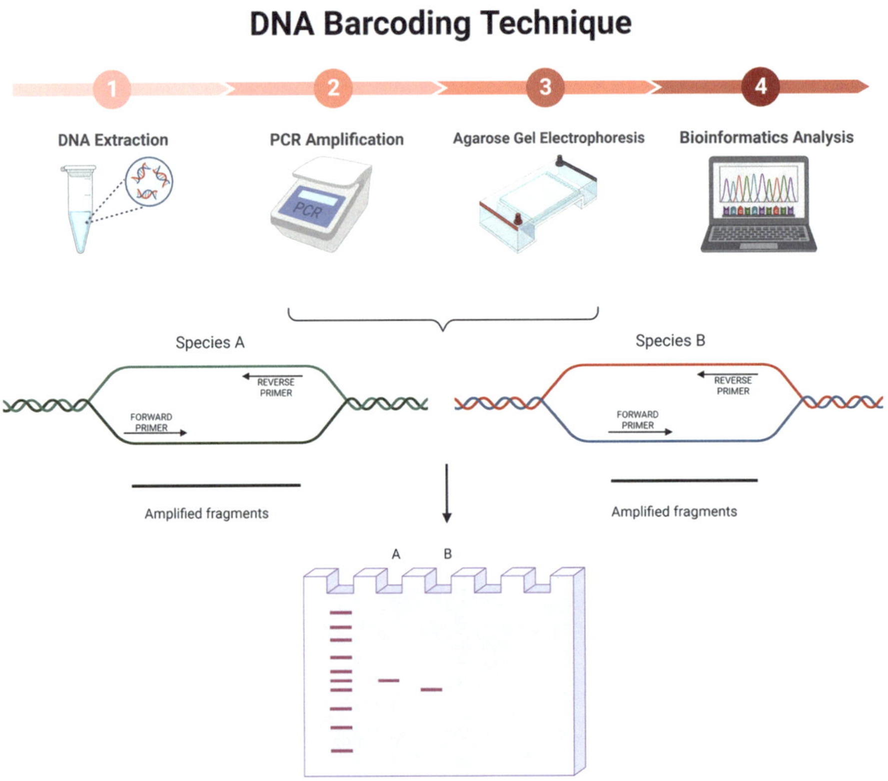

Figure 2.
DNA barcoding technique. Top: 1) DNA extraction and quantification. 2) amplification of a short DNA sequence by PCR with previously designed primers. 3) visualization of the amplified fragment by agarose gel electrophoresis. 4) Bioinformatic analysis of short DNA sequences or 'barcodes' and identification of the sample. Bottom: Schematic representation of the barcode amplification and visualization after agarose gel electrophoresis from species A and B. Created with Biorender.com.

specific fragment of the expected size is obtained, the amplification product is sequenced by Sanger for subsequent bioinformatic analysis and taxonomic identification (**Figure 2**).

2.1.1 Extraction and quantification of genomic DNA

After sample collection and photographic documentation, DNA is extracted using standardized protocols or specific kits, which yield good quality, inhibitor-free genomic DNA for PCR amplification. Tissue lysis is performed with Proteinase K to degrade proteins such as nucleases that could degrade DNA during purification.

For the extraction of genomic DNA from animals, the use of the Edwards Buffer (200 mM Tris–HCl, pH = 7.5; 200 mM NaCl; 25 mM EDTA, 0.5% SDS; 0.1% β- Mercaptoetanol 1 μl/ml) [10, 11] is recommended. Because the high amount of phenolic compounds in plants, the DNA extracted with the Edwards Buffer protocol sometimes contains numerous PCR inhibitors, and the use of the commercial DNeasy® Plant ProKit (Qiagen N.V, Hilden, Germany) or CTAB buffer (0.1 M Tris–HCl (pH 8.0), 1.4 M NaCl, 0.02 M EDTA, 0.2 g/mL cetyltrimethyl-ammonium-bromide) (CTAB) [11, 12] is recommended.

DNA is quantified by spectrophotometer (NanoDrop ND-1000) and the ratio of absorbances 260:280 and 260:230 is used as a measure of the quality of DNA extraction. The former provides information on the purity of DNA and RNA in the sample: a ratio of ~1.8 is considered as acceptable for DNA analysis and a ratio of ~2.0 is accepted as acceptable for RNA. If compounds such as proteins, phenols or other contaminants that absorb at 280 nm are present in the sample, this ratio will be lower. Similarly, the presence of contaminants that absorb at 230 nm, such as carbohydrates, guanidine HCl or phenolic solutions [13], will result in values below ~2.0 in the 260:230 ratio.

2.1.2 Design and selection of primers

Table 1 shows the different primers selected and tested for different taxonomic groups. Although they are functional for many taxa, for certain groups it is necessary to design specific primers that allow the amplification of the desired sequences [14]. **Table 2** shows the PCR conditions for each pair of primers.

2.1.3 PCR amplification

We have performed PCR amplifications in 50 μl of reaction volume in the presence of 1.25 units of GoTaq® G2 Flexi DNA polymerase 5 u/μl (Promega, Wisconsin, USA), 0.2 mM each dNTP, 0.2 μM of each primer and less of 0.2 μg of template DNA.

The PCR amplification program depends on the primers chosen (**Table 2**). PCR reactions were heated 5 min. at 94°C, followed by 35 cycles under the following conditions: 15–30 secs. at 94°C (Denaturation step); 15–40 secs. at 50–60°C (Annealing step), 1–3 min. at 72°C (Elongation step).

2.1.4 Agarose gel electrophoresis

Samples are analyzed by standard agarose gel electrophoresis in 1% agarose in TBE buffer run at 3 V/cm at room temperature for 60–90′. Gels are prestained with 0.5 mg/ml ethidium bromide or GreenSafe© 1:20000. The HyperLadder 1 kb (Meridian Bioscience, Ohio, USA) or the Invitrogen DNA 1 kb Plus Ladder (Thermo Fisher Scientific, Massachusetts, USA) were used as molecular weight marker. A sample is shown in **Figure 3**.

PCR program	Initial step	Number of Cycles	Denaturation step (A)	Annealing step (B)	Extension step (C)	Additional extension step
(1)	94°C 4′	35	94°C 1′	54°C 30′	72°C 1′	—
(2)	94°C 4′	35	94°C 15′	54°C 15′	72°C 30′	—
(3)	94°C 3′	35	94°C 30′	48°C 40′	72°C 1′	72°C 10′
(4)	95°C 2,5′	35	95°C 30′	56°C 30′	72°C 45′	72°C 10′
(5)	94°C 3′	35	94°C 30′	52°C 30′	72°C 45′	72°C 10′

Table 2.
PCR protocol for the amplification of the different group of organisms described in **Table 1**. *Modified from [6].*

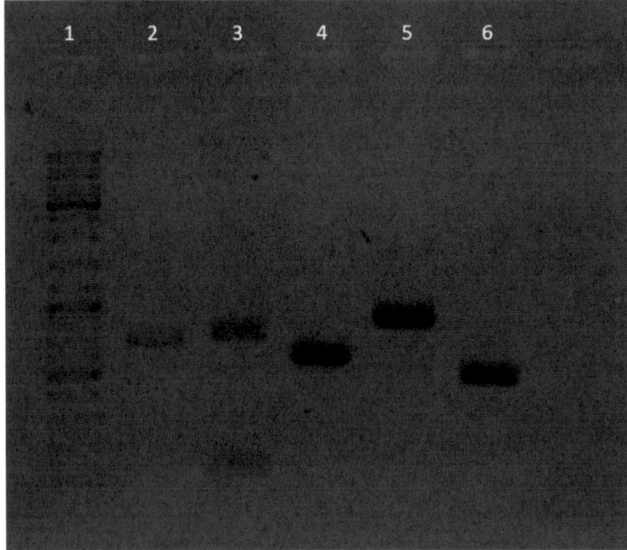

Figure 3.
Analysis by agarose gel electrophoresis of the PCR products obtained for DNA barcoding analysis. Reactions were performed as described in the text. Lane 1: HyperLadder 1 kb. Lane 2: amplification of Gibbula divaricata COI barcode using LCO1490 and HCO2198 primers specific for invertebrates. Lane 3: amplification of Lophius vomerinus COI barcode using FISHCOILBC_ ts and FISHCOIHBC_ts primers specific for fish. Lanes 4, 5, 6: amplification of barcodes rbcL (rbcLa f and rbcLa rev primers), matK (matk-3F and matk-1R primers) and nrITS2 (nrITS2-S2F and nrITS2-S3 Rev. primers) respectively from the plant Nerium oleander.

2.1.5 DNA sequencing and bioinformatic analysis

PCR products are purified to remove components of the PCR reactions, resuspended in TE buffer and reverse and direct sequences are obtained by standard Sanger sequencing. To identify the species to which the obtained sequence corresponds, bioinformatic analysis involves the following steps: (1) Data preprocessing: quality control of the raw sequence data (.ab1 file format), which may include filtering out low-quality reads, trimming the end of the sequences, and removing any artifacts or contaminants. Free licensed software such as SnapGene Viewer can be used for this purpose. (2) Sequence alignment: preprocessed DNA sequences are aligned to a reference database. Alignment algorithms such as BLAST (Basic Local Alignment Search Tool) are used. Identification is performed based on sequence similarity. (3) Data analysis, which may include the calculation of genetic distances and phylogenetic trees.

It is particularly interesting to work with specific databases such as the Barcode of Life Data Systems (BOLD, www.boldsystems.org), which provides a repository of DNA barcode sequences of various species using the DNA Barcoding technique [14]. BoldSystem compiles all the information necessary for the taxonomic identification of an individual, short DNA sequence or "barcode", taxonomic descriptions, photographs of specimens, geolocation and the sequences of other individuals of the species. The database itself, when including the problem sequence/s in the system, will return a phylogenetic tree based on the Neighbor-Joining statistical method, in which the species can be included within the corresponding phylogenetic group by comparing it with the sequences collected in the database. Furthermore, the phylogenetic tree obtained clearly includes the locations of all the sequences of the species with which the problem sequence has been identified, making it possible to study intraspecific diversity by inferring the phylogeographic character of the different variants of a species. This process is of great importance for studies of biogeography, invasive alien species or population ecology [15].

3. Random amplified polymorphic DNA and uses

Random Amplification of Polymorphic DNA (RAPD) is a molecular biology technique that can be used for genetic analysis and identification of organisms when DNA barcoding is not feasible due to the absence of reference sequences in the databases or because the genetic proximity of species offers little genetic diversity in the genes used as barcodes [16]. RAPD is a PCR-based technique that amplifies random segments of DNA using arbitrary short primers. It requires no prior knowledge of the DNA sequence and can generate a unique, fingerprint-like pattern of amplified DNA fragments for each individual or species [17, 18]. RAPD analysis relies on the presence or absence of amplified DNA fragments, and the resulting banding patterns can be visualized by gel electrophoresis.

RAPD is a relatively simple, fast and low-cost technique, so it is used in several fields of genetics. The main advantage is that it is not necessary to know the genetic sequence to be studied, so when the starting information is insufficient to create specific primers, it is one of the best options for genetic studies. Other advantages are the possibility of genotyping a wide variety of organisms, from bacteria to animals or plants using the same primers and the high number of polymorphisms obtained in each test [18–21].

3.1 Random amplified polymorphic DNA technique

The use of random sequence primers was first described in 1990 by Welsh and McClelland in their bacterial genome assay "Fingerprinting genomes using PCR with arbitrary primers", which they called "Arbitrarily Primed Polymerase Chain Reaction" (AP-PC) [19, 22]. In the same year, Williams et al. published another paper describing a similar technique called "randomly amplified polymorphic DNA" (RAPD) [17, 19]. These two techniques, together with DNA amplification fingerprinting (DAF), have been grouped under the term Multiple Arbitrary Amplicon Profiling (MAAP) [20, 21]. All are based on the polymerase chain reaction (PCR) and follow the same principle: random amplification of DNA fragments, the main difference being the length of the primers, the annealing temperature and the type of gel electrophoresis used to visualize the bands generated [17, 20, 22, 23], with RAPD being the most popular due to its simplicity and low cost [23].

RAPDs use a single primer per PCR reaction, as opposed to the two primers used in traditional PCR. The primer has a length of 10 nucleotides and an arbitrary sequence with a GC content equal to or greater than 60% [17, 21] and acts as a direct and reverse primer at the same time (**Figure 4**). Using this type of primer, it is possible to find numerous loci to bind to throughout the genome [18, 19, 24]. Furthermore, in this type of PCR, a low annealing temperature is set so that, even if the complementarity of the primer with the strand is not perfect, around 90%, as long as the last bases of the 3′ end match, binding will occur, thus increasing the number of binding sites in the genome [19]. However, only fragments in which the binding site of one strand is between 0.5 and 4 kilobases from the binding site of the complementary strand and the 3′ ends are facing each other will be amplified (**Figure 4**) [18, 19, 24].

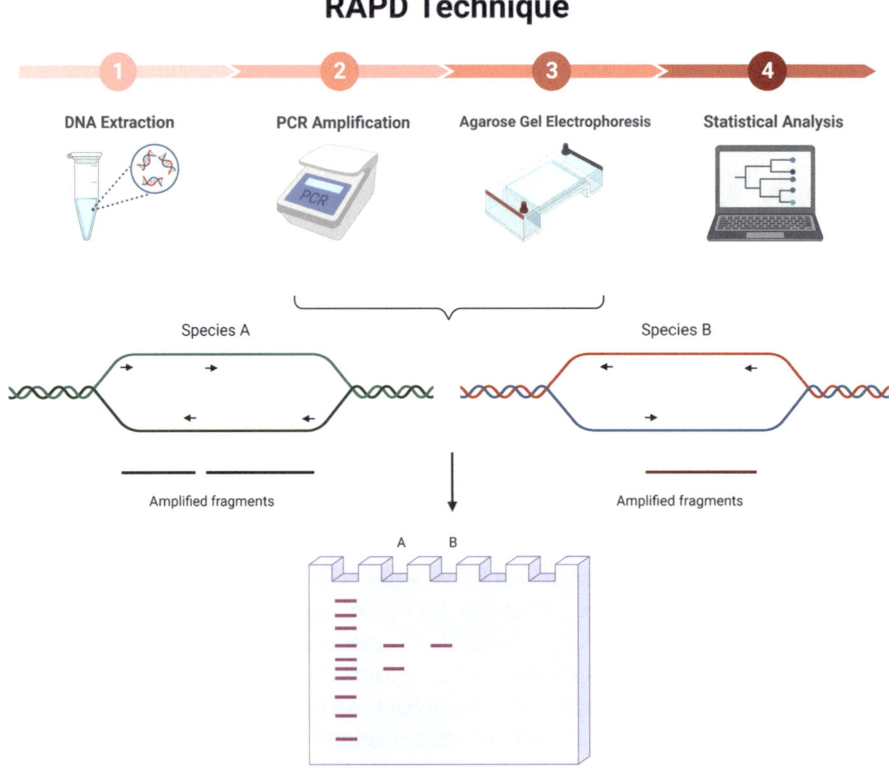

Figure 4.
RAPD technique involves 4 steps. Top: (1) sample DNA extraction, (2) PCR amplification with a specific primer, (3) agarose gel electrophoresis and, (4) statistical analysis. Bottom: Primer hybridizes at random along the genome, rendering two amplified bands in the case of species A and just one band in species B. results are visualized after agarose gel electrophoresis. Created with Biorender.com.

When using the RAPD technique, it must be assumed that the bands of the different labeled loci do not migrate the same distance on the gel [16]. However, it may happen that among the amplified products bands with the same molecular weight are found in different species, which could be due to homologous traits, those inherited from a common ancestor, or homoplastic traits, those that have arisen independently in a population. It follows that if two individuals have the same band, their degree of relatedness is higher [24].

With the data obtained, a statistical study can be carried out by generating a binary matrix in which a value of 1 is given to the presence of a certain band in the gel and 0 for its absence, in order to subsequently calculate similarity coefficients of the individuals, allelic frequency, distance matrices or generate dendrograms using various computer programs [20, 25–27].

3.2 Amplification conditions

We have performed PCR amplifications for RAPDs in 50 μl of reaction volume in the presence of 1.25 units of GoTaq® G2 Flexi DNA polymerase 5 u/μl (Promega, Wisconsin, USA), 0.2 mM each dNTP, 0.4 μM of each primer and less of 0.5 μg of template DNA.

PCR amplification conditions must be set up considering the type of primers used. It is recommended to optimize the conditions by conducting a gradient PCR or varying the annealing temperature, primer concentration, or other parameters to achieve the best results and reproducibility of the specific experiment (**Figure 4**), [18, 19]. Low annealing temperature promotes the appearance of primer-dimer formation. On the contrary, high annealing temperature reduces the number of polymorphic bands. PCR reactions are performed with just one primer instead of two. Samples were heated three or four min. at 94–95°C, followed by 35–45 cycles under the following conditions: 30–60 s at 92–94°C (Denaturation step); 40–120 s at 36–40°C (Annealing step), 90–120 s at 72°C (Elongation step), and a final extension at 72°C for 7–10 min. [27–30]. We achieved the best results with the following conditions: 4 min. at 94°C, followed by 40 cycles of 40 s at 94°C, annealing at 40°C for 45 s, and elongation at 72°C for 90 s, and a final extension at 72°C for 10 min.

3.3 Agarose gel electrophoresis

After PCR, the banding pattern generated is visualized on an agarose gel using a classical electrophoresis apparatus. For each sample, 3 to 20 polymorphic bands can be obtained, with molecular weights ranging from 0.5 to 5 kilobases, depending on the distance between the primer binding sites (**Figure 4**) [18, 19]. Due to the large number of bands obtained and the variety of sizes, the gel must be run at low voltage [27, 30]. The best results in our laboratory have been obtained using a voltage of 4.5 V/cm^2 for 120 minutes in a medium-sized (15 × 10 cm) tray. Differences can be obtained by varying the PCR amplification conditions (see **Figure 5**).

Gels are prepared at a concentration of 1.5–2.0% agarose, as this favors the separation of smaller fragments [20, 27, 30, 31]. The intercalating agent we prefer to use is ethidium bromide for RAPD, as it shows better results than other alternatives such as SYBR™ Safe or GreenSafe [23, 25], but certain safety measures must be followed because of its mutagenic effect. Alternatively, a polyacrylamide gel with AgNO3 as a stain can be used, as is done in the AP-PCR technique [16, 19, 22, 23].

Although RAPD has limitations, such as difficulties in reproducibility and difficulties in interpreting complex banding patterns, it can provide valuable insights into the genetic makeup and relationships of organisms when other methods like DNA barcoding are not feasible [18–21].

RAPD can be used in conjunction with other molecular markers, such as microsatellites, to obtain a more complete picture of genetic diversity to identify individuals within a species. Microsatellite typing, also known as simple sequence repeat (SSR) typing, is a molecular biology technique used to genotype individuals based on short tandem repeat (STR) or microsatellite variations in the DNA. Microsatellites are short DNA sequences, usually consisting of 1–6 base pair motifs, which are tandem repeats and are highly variable in length and number of repeats. The instability of microsatellite repeats occurs because of slippage errors in replication [32] or defects in repair by the mismatch repair machinery, leading to alterations in the number of repeat units. During DNA replication, a repetitive DNA sequence, such as a microsatellite, may form a hairpin loop formed by the repetitive sequence itself. These secondary structures can cause DNA polymerase to stall and dissociate from the template strand, leading to the slippage error [32–34]. These alterations consist in microsatellite expansions (increased number of repeat units) and contractions (decreased number of repeat units).

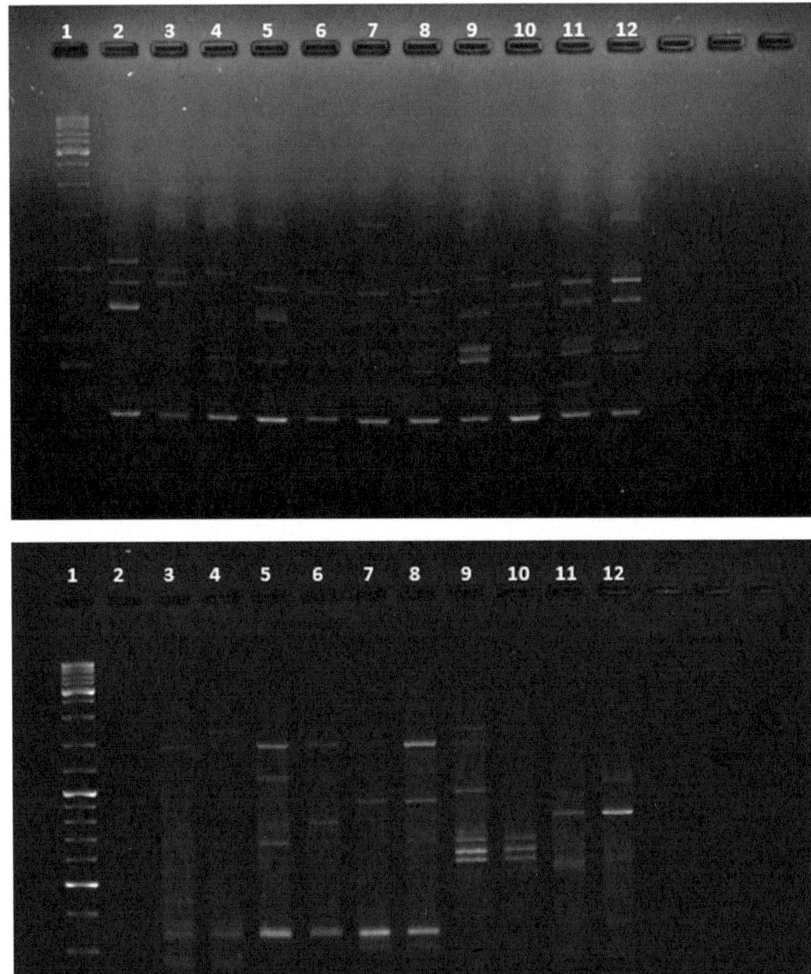

Figure 5.
Analysis by agarose gel electrophoresis of the RAPDs products, the OPD-11 (5'- AGCGCCATTG - 3') primer was used. A comparison of amplification profiles by RAPD technique is shown. Top: Reaction conditions: 45 cycles and annealing temperature at 36°C. Bottom: The number of cycles was decreased to 40 and the annealing temperature was increased to 40°C. Lane 1: molecular marker. Lane 2: negative control (mixture PCR + water). Lanes 3, 4: Quercus coccifera. Lanes 5, 6: Quercus faginea. Lanes 7, 8: Quercus robur. Lanes 9, 10: Quercus rotundifolia. Lanes 11, 12: Quercus suber. The primer-dimer phenomenon can be observed on top gel as a common band that migrates ahead on the gel. This band disappears in the gel at botom when changing the PCR conditions and achieving, in part, more specific binding.

Used to genotype individuals based on variations in STRs in DNA, it has been widely used in population genetics, forensic science, and plant and animal breeding. However, it has the limitation that the DNA sequence flanking the target microsatellite must be known for primer design. As for RAPDS, SSR markers only indicate the presence or absence of amplified fragments in a sample, so they considered dominant markers because they do not provide information on heterozygosity or allelic dosage [18, 21, 23].

4. Comparison with other methods

In this work we have used DNA Barcodes as well as RAPDs for the identification of species. Other methods are based on enzymatic digestion (RFLP, Restriction fragment length polymorphism), PCR (ISSR, Inter-Simple Sequence Repeat), or combination of both procedures (AFLP, Amplified fragment length polymorphism). The latter two proposals solve problems of RAPD assays such as reproducibility and specificity. In the case of ISSR, this is due to the longer primer length and higher PCR annealing temperature, despite the use of a random sequence primer. However, they cannot distinguish between homozygotes and heterozygotes [18, 19, 21, 23]. In addition, the cost of the AFLP technique is quite high and requires a high level of expertise because the procedures are more complex [21]. RFLP can be used to distinguish heterozygosity due to its codominant character; however, as it requires a large amount of starting DNA, knowledge of the sequence to be studied and because is not very useful for differentiating between evolutionarily close species, it is not a widely used alternative [18, 19, 21, 23].

5. Conclusions

The use of barcoding and RAPDs (Random Amplified Polymorphic DNA) techniques has significantly contributed to the identification of species in various fields of research, including biology, ecology, conservation, and forensics. Both methods offer valuable tools for species identification, each with its own strengths and limitations.

Barcoding, which involves the analysis of specific DNA regions, such as the mitochondrial cytochrome c oxidase subunit I (COI) gene or other barcodes, has gained widespread recognition as a powerful tool for species identification. The barcodes described in the text exhibit sufficient variation across species, allowing for reliable differentiation even among closely related taxa. Barcoding offers several advantages, including its simplicity, cost-effectiveness, and the ability to identify species based on fragmented or degraded DNA samples. This technique has been particularly useful in large-scale biodiversity assessments, tracking invasive species, and detecting species in complex environmental samples. However, studies on species hybridization have shown certain limitations, so RAPDs have been used in these cases. These molecular markers have also been widely used for population genetics studies, phylogenetic analysis, and identification of closely related species. The technique is relatively simple and cost-effective, requiring minimal DNA sequence information. However, RAPDs have some limitations, such as issues with reproducibility, sensitivity to DNA quality and quantity, and difficulties in standardization and comparison across different laboratories.

The combined use of barcoding and RAPDs has proven to be a powerful approach, particularly when addressing complex taxonomic or evolutionary questions. By integrating the information provided by both techniques, researchers can enhance the accuracy and resolution of species identification. Barcoding can provide a reliable species-level identification, while RAPDs can reveal intraspecific genetic variation and patterns, allowing for the study of population structure and differentiation. The continued advancement of sequencing technologies and bioinformatics tools will further refine and expand the application of these techniques, enabling more efficient and accurate species identification in the future.

Acknowledgements

Part of this work was funded by the project OpenLab (SmartCampus, University of Malaga, Spain).

Author details

Estefanía García-Luque, Ana del Pino-Pérez and Enrique Viguera*
University of Malaga, Malaga, Spain

*Address all correspondence to: eviguera@uma.es

References

[1] Hebert PDN, Cywinska A, Ball SL, de Waard JR. Biological identifications through DNA barcodes. Proceedings of the Royal Society of London Series B: Biological Sciences. 2003;**270**(1512):312-321. DOI: 10.1098/rspb.2002.2218

[2] Trivedi S, Rehman H, Saggu S, Panneerselvam C, Ghosh SK, editors. DNA Barcoding and Molecular Phylogeny. Switzerland: Springer International Publishing; 2020. DOI: 10.1007/978-3-030-50075-7

[3] Shadrin DM. DNA barcoding: Applications. Russian Journal of Genetics. 2021;**57**(4):489-497. DOI: 10.1134/S102279542104013X

[4] Kress WJ. Plant DNA barcodes: Applications today and in the future: Plant DNA barcode applications. Journal of Systematics and Evolution. 2017;**55**(4):291-307. DOI: 10.1111/jse.12254

[5] Gouy M, Guindon S, Gascuel O. SeaView version 4: A multiplatform graphical user interface for sequence alignment and phylogenetic tree building. Molecular Biology and Evolution. 2010;**27**(2):221-224

[6] Cold Spring Harbor Laboratory DNA Learning Center. Using DNA Barcodes to Identify and Classify Living Things [Internet]. 2018. Available from: https://dnabarcoding101.org/files/using-dna-barcodes.pdf [Accessed: May 01, 2023]

[7] CBOL Plant Working Group. A DNA barcode for land plants. Proceedings of the National Academy of Sciences. 2009;**106**(31):12794-12797. DOI: 10.1073/pnas.0905845106

[8] de Vere N, Rich TC, Trinder SA, Long C. DNA barcoding for plants. Methods in Molecular Biology. 2015;**1245**:101-118. DOI: 10.1007/978-1-4939-1966-6_8

[9] Schoch CL, Seifert KA, Huhndorf S, Robert V, Spouge JL, Levesque CA, et al. Nuclear ribosomal internal transcribed spacer (ITS) region as a universal DNA barcode marker for Fungi. Proceedings of the National Academy of Sciences of the United States of America. 2012;**109**(16):6241-6246. DOI: 10.1073/pnas.1117018109

[10] Majaneva M, Diserud OH, Eagle SHC, Hajibabaei M, Ekrem T. Choice of DNA extraction method affects DNA metabarcoding of unsorted invertebrate bulk samples. Metabarcoding and Metagenomics. 2018;**2**(2):1-12. DOI: 10.3897/mbmg.2.26664

[11] Whitlock R, Hipperson H, Mannarelli M, Burke T. A high-throughput protocol for extracting high-purity genomic DNA from plants and animals. Molecular Ecology Resources. 2008;**8**(4):736-741. DOI: 10.1111/j.1755-0998.2007.02074.x

[12] Tamari F, Hinkley CS, Ramprashad N. A comparison of DNA extraction methods using Petunia hybrida tissues. Journal of Biomolecular Techniques JBT. 2013;**24**(3):113-118. DOI: 10.7171/jbt.13-2403-001

[13] Thermo Scientific. 260/280 and 260/230 Ratios [Internet]. 2019. Available from: https://dna.uga.edu/wp-content/uploads/sites/51/2019/02/Note-on-the-260_280-and-260_230-Ratios.pdf [Accessed: May 01, 2023]

[14] Ratnasingham S, Hebert PDN. Bold: The barcode of life data system.

Molecular Ecology Notes. 2007;**7**(3):355-364. DOI: 10.1111/j.1471-8286.2007.01678.x

[15] Gostel MR, Kress WJ. The expanding role of DNA barcodes: Indispensable tools for ecology, evolution, and conservation. Diversity. 2022;**14**(3):213. DOI: 10.3390/d14030213

[16] Alcántara MR. Breve Revisión de Marcadores Moleculares. In: Eguiarte LE, Souza V, Aguirre X, editors. Ecología molecular. Secretaría de Medio Ambiente y Recursos Naturales, Instituto Nacional de Ecología; 2007. pp. 541-566

[17] Williams JGK, Kubelik AR, Livak KJ, Rafalski JA, Tingey SV. DNA polymorphisms amplified by arbitrary primers are useful as genetic markers. Nucleic Acids Research. 1990;**18**(22):6531-6535. DOI: 10.1093/nar/18.22.6531

[18] Kumar P, Gupta VK, Misra AK, Modi DR, Pandey BK. Potential of molecular markers in plant biotechnology. Plant Omics Journal. 2009;**2**:141-162

[19] Liu Z. Randomly amplified polymorphic DNA (RAPD). In: Liu Z, editor. Aquaculture Genome Technologies. 2007. p. 21-28. DOI: 10.1002/9780470277560.ch3.

[20] Wojciechowska-Koszko I, Mnichowska-Polanowska M, Roszkowska P, Sławiński M, Giedrys-Kalemba S, Dołęgowska B, et al. Improved RAPD method for Candida parapsilosis fingerprinting. Genes. 2023;**14**(4):868. DOI: 10.3390/genes14040868

[21] Grover A, Sharma PC. Development and use of molecular markers: Past and present. Critical Reviews in Biotechnology. 2016;**36**(2):290-302. DOI: 10.3109/07388551.2014.959891

[22] Welsh J, McClelland M. Fingerprinting genomes using PCR with arbitrary primers. Nucleic Acids Research. 1990;**18**(24):7213-7218. DOI: 10.1093/nar/18.24.7213

[23] Idrees M, Irshad M. Molecular markers in plants for analysis of genetic diversity: A review. European Academic Research. 2014;**2**:1513-1540

[24] Tingey SV, del Tufo JP. Genetic analysis with random amplified polymorphic DNA markers. Plant Physiology. 1993;**101**(2):349-352. DOI: 10.1104/pp.101.2.349

[25] Kumari. Randomly amplified polymorphic DNA – A brief review. American Journal of Animal and Veterinary Sciences. 2014;**9**(1):6-13. DOI: 10.3844/ajavsp.2014.6.13

[26] Dwivedi S, Singh S, Chauhan UK, Tiwari MK. Inter and intraspecific genetic diversity (RAPD) among three most frequent species of macrofungi (*Ganoderma lucidum*, *Leucoagricus* sp. and *Lentinus* sp.) of tropical forest of Central India. Journal of Genetic Engineering and Biotechnoly. 2018;**16**(1):133-141. DOI: 10.1016/j.jgeb.2017.11.008

[27] Ardi M. Genetic variation among Iranian oaks (Quercus spp.) using random amplified polymorphic DNA (RAPD) markers. African Journal of Biotechnoly. 2012;**11**(45):10291-10296. DOI: 10.5897/AJB12.325

[28] Beteş D, Çolak R, Karacan GO, Kandemir İ, Kankiliç T, Çolak E. Population genetic variability of Myodes glareolus (Schreber, 1780) (Mammalia: Rodentia) distributed in Northern Anatolia as revealed by RAPD – PCR

analysis. Acta Zoologica Bulgarica. 2014;**66**(1):31-37

[29] Barreneche T, Bodenes C, Lexer C, Trontin JF, Fluch S, Streiff R, et al. A genetic linkage map of *Quercus robur* L. (pedunculate oak) based on RAPD, SCAR, microsatellite, minisatellite, isozyme and 5S rDNA markers. Theoretical and Applied Genetics. 1998;**97**(7):1090-1103. DOI: 10.1007/s001220050996

[30] Subositi D, Mursyanti E, Widodo H, Widiyastuti Y. RAPD primer screening for markers development of Jinten Hitam (*Nigella sativa* L.) authentication. Rubiyo, Indrawanto C, editors. E3S Web of Conferences. 2021;**306**:01008. DOI: 10.1051/e3sconf/202130601008

[31] Ylmaz M, Ozic C, Gok L. Principles of nucleic acid separation by agarose gel electrophoresis. In: Magdeldin S, editor. Gel Electrophoresis – Principles and Basics. London, UK: InTech; 2012. DOI: 10.5772/38654

[32] Viguera E, Canceill D, Ehrlich SD. Replication slippage involves DNA polymerase pausing and dissociation. EMBO Journal. 2001;**20**(10):2587-2595. DOI: 10.1093/emboj/20.10.2587

[33] Viguera E, Canceill D, Ehrlich SD. In vitro replication slippage by DNA polymerases from thermophilic organisms. Journal of Molecular Biology. 2001;**312**(2):323-333. DOI: 10.1006/jmbi.2001.4943

[34] Castillo MG, Henneke G, Viguera E. Replication slippage of the thermophilic DNA polymerases B and D from the Euryarchaeota Pyrococcus abyssi. Frontiers in Microbiology. 2014;**5**:403. DOI: 10.3389/fmicb.2014.00403

Chapter 5

The Application of Electrophoresis in Soil Research

Cheuk-Hin Law, Long-Yiu Chan, Tsz-Yan Chan, Yee-Shan Ku and Hon-Ming Lam

Abstract

Soil is a complex mixture of minerals and organic matters in which microbes, plants, and animals interact. In the natural environment, soil constantly undergoes physical, chemical, and biological transformations under the influences of environmental factors such as humidity and temperature. Studies on soil chemical compositions, microbes, and abundances of plants and animals provide useful information on the soil property for proper land use planning. Since soil is a complex mixture, soil studies require the effective separation of its various components, which can be achieved with electrophoresis, a powerful method that exploits the inherent differences in the physical and chemical properties of these components. By combining electrophoresis with other technologies such as chromatography, mass spectrometry, polymerase chain reaction (PCR), and DNA sequencing, substances including humic acids, amino acids, environmental pollutants, nutrients, and microbial, plant, and animal DNA can be identified and quantified. In this chapter, the applications of different electrophoresis-based technologies will be discussed with respect to soil research, and their principles, advantages, and limitations will be addressed.

Keywords: soil, microbe, eDNA, plant-microbe interaction, animal-microbe interaction, electrophoresis, DNA marker, proteome

1. Introduction

The soil is a complex ecosystem in which inorganic and organic compounds, microbes, plants, and animals co-exist. The composition of soil is dynamic since it is heavily dependent on environmental conditions such as humidity and temperature. In the face of a changing climate, the soil composition fluctuates even more. Soil quality provides important information for the proper planning of land use, and is greatly influenced by microbial activities. For example, the degradation of soil organic matter (SOM) and the nitrogen-fixing and/or denitrifying activities of soil-borne microbes regulate the fluxes of carbon and nitrogen [1], which is important for the growth of all organisms in soil.

Electrophoresis is a versatile technique for studying the chemical and microbial compositions of soil, separating molecules based on their sizes and charges. To improve the resolution, variants of electrophoresis such as capillary electrophoresis,

IntechOpen

denaturing electrophoresis, and two-dimensional electrophoresis were developed. To further enhance the analytical capacity of electrophoresis, the technique has been coupled with other platforms, such as mass spectrometry and DNA sequencing, to reveal the identity of the molecules. In addition to resolving proteins and DNA, electrophoresis is also employed to prepare DNA extracted from soil for downstream analyses. The study of soil environmental DNA (eDNA) provides information on the identities and abundances of plants and animals in soil. Thus, electrophoresis has been demonstrated as a useful tool for soil research.

2. The application of capillary electrophoresis (CE) in soil chemical composition analyses

The chemical composition of soil includes soluble substances such as humic acids, amino acids, peptides, proteins, oligonucleotides, carbohydrates, pigments, toxins, pesticides, vitamins, and chiral compounds [2]. Capillary electrophoresis (CE) has been proven to be a powerful analytical technique for separating compounds within a sample [3]. The method was originally developed using glass tubes to separate solutes [4]. The advantage of CE is that it requires only a small amount of sample to produce a high-resolution output in a short amount of time. Due to its efficiency in separating different compounds, CE is widely used in forensic studies [5], biomedical sciences [6], and soil research [7].

2.1 General principle of CE

CE features the use of a capillary tube, usually with an inner diameter ranging from 20 to 200 μm, for analyte separation [8]. The small-diameter capillary enhances the resolution by reducing lateral sample diffusion and minimizing the temperature difference between the center and the wall of the capillary [8]. CE encompasses a diverse range of techniques that serve distinct analytical purposes. For instance, capillary zone electrophoresis separates analytes based on their charge-to-size ratios, while capillary isoelectric focusing (CIEF) separates analytes by exploiting their different isoelectric points. The experimental setup for each technique is tailored to suit its specific requirements. In the case of capillary zone electrophoresis, a fused silica capillary is employed to connect two electrolyte buffers, which are in turn connected to opposing electrodes [9]. Upon the application of voltage, the sample flows through the capillary and its components are separated based on their charge-to-size ratios [9]. Detection and identification of the separated components are achieved using techniques such as UV-absorbance, fluorescence, and mass spectrometry [10].

2.2 Applications of different CE techniques

2.2.1 Capillary zone electrophoresis for humic acid and amino acid analyses

In soil, the decomposition of plant and animal matters containing oxygen-containing functional groups, such as ketones and carboxyl groups, results in humic substances (HSs) [11]. The humic fingerprint of a soil provides information on the soil property, palaeo-history, and pedogenesis [12]. The humic fingerprints of different layers of soil in a particular site is also useful for reconstructing the site history [12]. CE is a powerful technique for HS separation based on their charge-to-size ratios. For example, capillary zone electrophoresis could be used to generate the humic acid fingerprints of different soils [13]. The analyses revealed that humic acids from different soils exhibited different migration patterns and UV absorption profiles, allowing for their differentiation.

Amino acids serve as the building blocks of proteins and perform a variety of critical biological functions. In soil, amino acids play an important role in the nitrogen cycle in ecosystems. CE coupled with laser-induced fluorescence (CE-LIF) is a powerful tool for the detection of amino acids in soil. This technique involves the excitation of amino acids with laser light to a higher energy level, followed by the detection of the different emission wavelengths from different amino acids [14]. Using this approach, 17 common amino acids can be effectively separated and detected in 12 minutes [15].

Traditionally, chromatographic techniques have been widely used to analyze amino acids in soil [16]. Compared to the traditional techniques, CE offers several advantages including simpler procedures and greater accuracy. In contrast to chromatographic techniques, the CE-LIF analysis of amino acids in soil samples does not require a desalting step to remove any contamination from inorganic ions [16]. CE also provides a better resolution in amino acid separation in soil sample analyses [17]. However, compared to certain chromatographic techniques such as HPLC, the maximum amount of sample that can be applied to CE is much lower. Such a low input limits the detection capability of CE, but this limitation can be mitigated by the combined use of LIF [15].

2.2.2 Capillary isoelectric focusing for humic acid analysis

Besides capillary zone electrophoresis, capillary isoelectric focusing (CIEF) has also been employed to analyze soil humic acids. This was first carried out in 1997, where three distinctive humic acid fractions were obtained [18]. To improve the resolution, Kovács and Posta modified the traditional CIEF approach by applying methyl cellulose (MeC) in the additives for electrophoresis, as well as adding polyvinyl alcohol (PVA) to protect the capillary coating [19]. The modified approach resulted in the separation of humic acids into 30–50 fractions [19].

2.2.3 Micellar electrokinetic chromatography for the separation of ionic and neutral compounds

Micellar electrokinetic chromatography (MEKC) is a modified form of capillary electrophoresis (CE) that offers several advantages over traditional CE techniques. MEKC employs micelles as the pseudo-stationary phase, instead of silica as in traditional CE. This is achieved by adding surfactants to the buffer solution at concentrations higher than the critical micellar concentration, which leads to the formation of micelles [20]. As a result, MEKC is capable of separating both ionic and neutral compounds [20], making it a versatile tool for the analysis of a wide range of analytes. MEKC was used in conjunction with preconcentration techniques to detect trace amounts of sulfonylurea herbicides in soil [21]. However, MEKC is not capable of separating larger molecules with molecular weights greater than 5 kDa [22], such as proteins and oligo-saccharides [23].

2.2.4 Microchip capillary electrophoresis for soil nutrient analysis

In studying soil composition, soil nutrient analysis is of particular importance, as it affects soil fertility, which in turn affects crop yield in agriculture. Changes in weather, water, and nutrient contents can have significant impacts on soil productivity [24]. By analyzing the nutrient content of soil, changes can be detected and addressed before they have a significant negative impact on the agricultural output [24]. CE can be used to study the inorganic nutrient contents of soil by separating ions with different sizes and charges, and the respective concentrations of these ions in the sample can then be determined by measuring their conductivity [25].

In one study, a mobile sensor was used to measure ion concentrations in water-extracted soil samples using microchip capillary electrophoresis (MCE) [25]. The ion concentrations of four different soil types, including Cambisol, Luvisol, Phaeozem, and Anthrosol, were successfully measured in the study. However, it should be noted that other ions, such as NH_4^+, K^+, and PO_4^{3-}, may require further testing with a wider range of concentrations in the sample.

2.2.5 The advantages and limitations of CE in soil research

One of the major advantages of MCE is its high sensitivity in measuring ions, allowing for the accurate and precise quantification of nutrient concentrations in soil samples [25]. Additionally, MCE requires only a simple aqueous extraction of the sample before analysis, reducing the need for costly and time-consuming sample preparation steps. Specifically, the sample only requires shaking with deionized water and filtering through a 0.22 μm syringe filter. This simple extraction method lowers the cost and complexity of sample preparation, making CE an attractive option for routine soil nutrient analysis, especially in the field. However, MCE has limitations regarding the separation of ions under high concentrations, so further validation of its functionality in measuring a wider range of soil nutrient concentrations is needed.

3. The application of differential gradient gel electrophoresis (DGGE) and two-dimensional polyacrylamide gel electrophoresis (2D-PAGE) in soil microbe studies

The soil microbial community consists mainly of millions of species of bacteria and fungi that perform diverse functions [26]. The microbes interact with one another as well as with plants and animals. For example, *Rhizobium spp.* interact with legume plants to form nitrogen-fixing nodules [27]. The diversity of soil microbes could also reflect invertebrate activities in the soil [28]. In addition, soil microbes decompose dead organisms in the soil to recycle the nutrients [29]. Studies on soil microbial communities will facilitate the understanding of the interactions among soil microbes, plants, and animals in the soil to promote ecological sustainability.

3.1 General principle of DGGE

DGGE was first utilized to test the *EcoRI* fragment of *λ* or *E. coli* DNA [30]. The technique has been widely employed to study microbial communities in various fields, such as medical science, marine science, and soil science [31–33]. DGGE is usually coupled with polymerase chain reaction (PCR), which amplifies marker genes such as 16S rDNA [34]. DGGE then generates a fingerprint of the amplicons for a specific microbial community.

Traditional electrophoresis separates analytes based on size differences but DGGE accomplishes this based on differences in nucleotide compositions [35]. Separation is achieved using a denaturing gel matrix for electrophoresis. Examples of denaturants include urea and formamide. By varying the concentrations of these denaturants along a gradient in the gel, different degrees of ease of strand separation can be established [33]. Compared to those with lower GC contents, DNA fragments with higher GC contents require more denaturant to achieve strand separation due to the stronger binding between G-C base pairs than A-T base pairs [36]. Since the 16S rDNA sequence varies in different microbes, the number and intensity of DNA bands observed on the gel can indicate the diversity and relative abundances, respectively, of a particular group of microbes in a soil sample [33].

3.1.1 The application of DGGE in studying soil microbial diversity

DGGE is widely used for microbial community analyses. It was first applied in ribosomal sequence analyses in a microbial community study [37]. The *16S rRNA* gene encodes the small subunit ribosomal RNA of prokaryotic ribosomes, which is involved in the translation of messenger RNAs (mRNAs) into proteins to perform various biological functions. 16S rRNA is frequently analyzed in phylogenetic studies due to its conserved features and its utility as a “molecular clock” for deducing the evolutionary divergence between two living entities [38]. For example, a study analyzed soil microbial communities from different regions subjected to various agricultural management practices, such as crop rotation, and also from contaminated soil [39]. In this study, PCR amplification was performed on specific sequences in the V3 and V6/V9 regions of *16S rRNA* gene in bacteria and the V3 region of *16S rRN*A gene in archaea prior to DGGE analyses. The V3 region was chosen due to sequence variations that could result in distinct banding patterns in DGGE [37]. The analyses of the resulting banding patterns revealed that polyaromatic hydrocarbon (PAH)-contaminated soils had fewer bands compared to uncontaminated soils, indicating a decrease in bacterial diversity [39].

In addition to analyzing treatment-induced changes in the soil community structure, DGGE is also valuable for studying the compositions of microbes in various habitats, such as forests, and oil-contaminated paddy soil [33]. This enables the comparison of community structures between different areas. Moreover, this technique can be employed in conjunction with samples from plant tissues. For instance, DGGE was used to investigate the diversities of actinomycetes in the soil and roots of various rice cultivars [34].

Besides its applications in studying changes in community structures across different habitats and plant tissues, DGGE also allows researchers to focus on specific microbes by selecting primers to amplify targeted genes. For instance, the *nifH* genes, which encode nitrogen fixation–related proteins, can be analyzed by DGGE to reveal the diversity in nitrogen-fixing bacteria in rhizopheric soil [34]. By excising and sequencing the DNA bands, researchers were able to identify a *Rhizobium* strain among the nitrogen-fixing bacteria in the soil sample. This technique has also been employed to analyze bacteria with ammonia monooxygenase (AMO) activity and to study the diversity of fungi using 18S rDNA and internal transcribed spacer (ITS) region sequences [40–42]. The coupling with different gene amplicons allows DGGE to be a versatile tool.

3.1.2 The advantages and limitations of DGGE in soil research

One of the key advantages of DGGE is the capacity to analyze complex microbial communities in soil, allowing for the rapid investigation and comparison of community structures across ecosystems [43]. Unlike culture-based methods, DGGE enables the direct analysis of microbial communities without the need for bacterial culturing to provide a more comprehensive understanding of the community. This is especially significant as traditional culture-based methods capture only a fraction of the community, typically around 1% [44]. In addition to the microbial diversity revealed by the DNA banding patterns, DGGE also allows the excision and purification of the DNA bands for sequencing to reveal the microbial identities [45].

However, one significant limitation is the challenge in differentiating complex communities. DGGE detection is limited to targets with genome numbers larger than 10^6 g^{-1} dry soil [46]. It is also difficult to distinguish between soil microbial communities where different species may be present in relatively similar proportions [39]. The low DNA-focusing capability of DGGE usually results in diffused bands, which weakens its capacity to differentiate between microbial communities [33, 39]. Moreover, the heterogeneity of the genes amplified for DGGE can result in multiple bands even with a pure microbial culture [39]. The multiple bands may then lead to the misinterpretation of results.

3.2 General principle of two-dimensional polyacrylamide gel electrophoresis (2D-PAGE)

The principle behind 2D-PAGE is to separate analytes in two dimensions based on two parameters. Such a two-dimensional separation allows for a higher resolution in the separation compared to using only one parameter, that is, a one-dimensional separation. 2D-PAGE was first presented in the 1970s [47, 48] to separate proteins according to their isoelectric points (first dimension) and molecular weights (second dimension). A similar concept was adopted to separate DNA fragments based on different lengths and GC contents [42]. The separated proteins or DNA fragments can then be subjected to further analyses, such as mass spectrometry for protein identification [49] or DNA sequencing [42].

3.2.1 The applications of 2D-PAGE in analyzing soil microbial diversity

3.2.1.1 The separation of soil proteins using 2D-PAGE

2D-PAGE is a widely used technique for separating proteins within a mixture, facilitating the investigation of protein expressions in specific cell types [49] and even metaproteomics [50]. In the study by Paul and Nair, 2D-PAGE was used in conjunction with MALDI-TOF/MS (matrix-assisted laser desorption/ionization coupled with time-of-flight mass spectrometry) to study the differential protein expression patterns of a plant growth–promoting rhizobacterium (PGPR), *Pseudomonas fluorescens* MSP-393, under salt shock [49]. The resulting peptide mass fingerprint revealed 22 differentially regulated proteins. 2D-PAGE also enables the large-scale identification of proteins from soil. For example, more than 800 protein spots could be detected from different types of soil in a metaproteomic analysis [51]. Using 2D-PAGE coupled with MALDI-TOF/MS, and by comparing the protein profiles among untreated, mineral fertilizer–treated, and organic manure–treated paddy soils, it was found that the long-term application of pig manure promoted the functional and structural diversity of soil microbes [52].

Other than soil microbial diversity, 2D-PAGE has also been used to investigate the functional relationships between soil and plant. In a study on the rhizospheric soil of a flowering broomrape, *Rehmannia glutinosa*, using 2D-PAGE, 103 protein spots were successfully separated from the rhizospheric soil samples [53]. By comparing the protein expression patterns across the 2-year timeframe of *R. glutinosa* monoculture, differentially expressed proteins associated with the monoculture were identified. Together with the results from root exudate analyses, it was suggested that the root exudates accumulated during monoculture changed the soil microbial ecology. This study demonstrated the usefulness of 2D-PAGE coupled with MALDI-TOF/MS in the study of microbe-plant interactions.

3.2.1.2 The separation of soil DNA using 2D-PAGE

In addition to facilitating the study of protein profiles of soil microbes, 2D-PAGE can also be used to determine the soil microbial community structure through DNA separation and identification. For instance, genetic markers such as the internal transcribed spacers (ITSs) of soil microbes were amplified by PCR and separated by 2D-PAGE to reveal the operational taxonomic units (OTUs) [42]. In the study, the first dimension of separation was based on fragment lengths using a non-denaturing polyacrylamide gel while the second dimension was based on the nucleotide composition through DGGE [42]. The separation of DNA fragments based on their lengths prior to DGGE enhanced the resolution of the separation, thus enabling the detection of bacterial communities with high degrees of structural similarity [42].

3.2.2 The advantages and limitations of 2D-PAGE in soil microbe studies

After 2D-PAGE, the analyzed proteins remain intact. Such a feature allows their subsequent identification through the use of mass spectrometry. Compared to using DGGE alone, the coupling of DGGE with a non-denaturing separation of DNA samples can improve the resolution. In a study analyzing the same soil sample using DGGE alone versus 2D-PAGE coupled with DGGE, ten times more operational taxonomic units could be identified with the addition of 2D-PAGE [42].

Despite these advantages, 2D-PAGE has several limitations, such as the difficulty in reproducing results due to inconsistencies in the pH gradient with ampholytes, as well as low extraction rates of proteins with transmembrane domains and difficulties in visualizing proteins with low abundances [54]. Additionally, as protein sizes increase, such as those larger than 200–500 kDa, the efficiency of separating them on the polyacrylamide gel decreases, due to the shrinking differences in size on a logarithmic scale [30, 55].

4. The application of agarose gel electrophoresis, DGGE, and CE for soil environmental DNA analyses

Environmental DNA (eDNA) is a pool of total DNA extracted from environmental samples [56], and bulk samples containing mainly the organisms of a target taxon such as insects collected from traps [57]. eDNA is composed of mostly intracellular microbial DNA and extracellular DNA left by organisms through processes such as shedding and defecation [58]. In addition to the existence of microbes, eDNA also contains those of plants and animals that either live in or pass through that area where the sample is collected. It has been demonstrated that soil eDNA can provide rich information about the diverse community living in the terrestrial environment including microbes, plants, invertebrates, and mammals [41, 59–61].

As early as the 1980s, the presence and recovery of eDNA in different habitats, such as soil, sediment, and aquatic environment, was reported [62–64]. Distinct from traditional DNA sampling methods, eDNA is extracted directly from environmental samples instead of isolated target organisms, so it is a versatile, easy-to-perform, and non-invasive approach to collecting large-scale DNA samples of the organisms living in the habitat of interest [65].

4.1 The application of agarose gel electrophoresis in eDNA preparation and analysis

While many types of electrophoresis have been introduced in the previous sections, conventional agarose gel electrophoresis, which is a simple and effective way to separate DNA fragments, is particularly suited to soil eDNA preparation.

4.1.1 The general principle of agarose gel electrophoresis

Agarose gel electrophoresis separates DNA samples based on their lengths. Upon gelation, agarose polymers will non-covalently link with one another to form a porous network that allows DNA molecules to pass through during electrophoresis [66]. When DNA is loaded onto the agarose gel and an electric current is applied, negatively charged DNA will migrate towards the positive pole [66]. Since DNA has a relatively uniform mass-to-charge ratio, the rate of migration is mainly determined by the size of DNA [67]. Thus, DNA fragments of different molecular weights can be separated by different migration rates and distances traveled. The sizes of sample DNA can also be estimated by simultaneously running a DNA marker ladder with fragments of known sizes.

4.1.2 The applications of agarose gel electrophoresis in eDNA preparation

In addition to analyzing DNA samples based on the molecular weight, agarose gel electrophoresis is also used for soil eDNA preparation. Humic acids are often co-extracted during soil eDNA extraction. These polyphenolic compounds can interfere with enzymatic reactions including PCR and restriction digestion, reducing the utility of the extracted DNA, so it is essential to include a purification step after extraction [68]. A method known as in-gel patch electrophoresis was developed to remove humic acids by packing a chromatography patch inside an agarose gel [69]. To remove humic acids having higher molecular weights than the DNA, the sample is first applied onto the chromatography patch for electrophoresis. After some time, the electric current is reversed. At this stage, the humic acids are retained in the chromatography patch while the DNA migrates back into the agarose gel [69]. To remove humic substances with lower molecular weights than the DNA, the samples are subjected to electrophoresis until the humic acids are separated from the DNA and have migrated out of the agarose gel [69]. Compared to other techniques, including cell lysis followed by ion-exchange chromatography or conventional gel electrophoresis together with gel extraction, in-gel patch electrophoresis yields DNA with a higher purity as indicated by higher $A_{260/280}$ and $A_{260/230}$ ratios and more complete digestion by restriction enzymes [69, 70]. Agarose gel electrophoresis is also used to purify high molecular weight (HMW) DNA from crude extracts. The crude extract, loaded onto a 1% agarose gel, is subjected to electrophoresis at 20 V for 16 hours. After that, the separated HMW DNA can be purified from the gel and ligated into desired vectors for eDNA library construction [71]. This method has been used to construct soil eDNA libraries in studies investigating the access to natural product gene clusters and identifying bacterial tryptophan dimer gene clusters [72, 73].

4.1.3 The applications of agarose gel electrophoresis in eDNA quality assessment

eDNA extraction usually involves multiple steps such as grinding, homogenization, sonication, and cell lysis with sodium dodecyl sulfate (SDS). These steps may result in unintended DNA fragmentation. This shows up as smears in agarose gel electrophoresis [74]. Various eDNA extraction methods have thus been developed to avoid such mishaps. For example, the slow-stirring method yields high-purity eDNA with minimal DNA fragmentation, as visualized by agarose gel electrophoresis as a quality control step [75].

Agarose gel electrophoresis can be coupled with restriction digestion to assess the purity of the extracted eDNA. Many of the impurities in eDNA co-extracted from soil inhibit enzymatic activities. To enhance the eDNA purity, the floatation method is used to separate the soil matrix into layers to remove enzyme-inhibiting substances prior to extraction steps [76]. The extracted DNA is then subjected to restriction digestion followed by agarose gel electrophoresis to visualize the digestion products. A complete DNA digestion would mean the successful removal of enzyme inhibitors [76]. Agarose gel electrophoresis is a simple quality control step in the eDNA purification process. However, it has to be coupled with other procedures such as restriction digestion to be effective.

4.2 The applications of DGGE and CE in eDNA analyses

Soil eDNA contains a variety of DNA from different organisms, including bacteria, fungi, plants, invertebrates, and mammals [77]. Purified eDNA can be used for PCR amplification of organism-specific marker genes to selectively amplify the DNA of the organisms of interest from the soil eDNA for downstream analyses. Besides the analyses of microbial communities in soil as mentioned in the previous sections, DGGE and CE can also be applied to investigate the ecological footprints of plants and animals in the soil eDNA. For example, PCR-DGGE has been used to analyze nematode communities in soil [41]. In the study, soil eDNA extracted from various agricultural fields was used for PCR amplification of the nematode 18S rDNA using nematode-specific primer sets followed by DGGE, to analyze the nematode community in these fields by examining the different DNA banding patterns [41].

After the amplification of the target DNA, CE can be applied to titrate the PCR products, check fragment lengths, and monitor primer dimers, before high-throughput sequencing. These workflows have been employed to study various communities in the soil, including plants, insects, earthworms, and other arthropods [60, 61, 78–80]. The identification of various soil communities forms an integral part of the ecological research in different habitats.

5. Conclusion

Electrophoresis separates molecules in soil, including minerals, organic compounds, proteins, and DNA, based on their sizes and charges. The composition of soil minerals and organic compounds is an important indicator of soil quality while the protein and DNA profiles generated by electrophoresis reveal the structural biodiversity of the soil communities. Since soil is rich in microbes, electrophoretic techniques are commonly applied to study the structural diversity in soil microbial communities. Coupling them with other platforms such as mass spectrometry and DNA sequencing to identify the individual protein or DNA fragments further enhances the utility of electrophoresis in soil research. Soil protein electrophoresis has been coupled with mass spectrometry to reveal the relationships between root exudates and soil microbial community structure. Such an association enables the study of microbe-plant interactions in soil. In addition, the analyses of soil eDNA using electrophoresis can reveal the different soil communities including insects, earthworms, and other arthropods, thus

Electrophoretic techniques	Examples of application	Advantages	Limitations	References
Denaturing gradient gel electrophoresis (DGGE)	Soil microbial community analysis	Direct and comprehensive analysis of microbial communities without the need for culturing	Blurred band patterns in analyzing complex communities	[33, 43]
Two-dimensional polyacrylamide gel electrophoresis (2D-PAGE)	Separation of proteins in mixtures, soil community analysis	Higher resolution than one-dimensional gel electrophoresis	Difficulties in reproducing results, difficult to separate large proteins well	[51, 54]
Capillary zone electrophoresis	Humic substance and amino acid analysis	Faster analysis, simpler preparation, and no need for desalting with CE-LIF, higher resolution than chromatographic techniques	Lower sensitivity than HPLC	[15, 17]
Isoelectric focusing	Separation of proteins in mixtures	High resolution, able to concentrate target protein	Long processing time	[81]
Micellar electrokinetic chromatography	Environmental pollutant analysis	Capable of separating both ionic and neutral compounds	Difficulties in separating large molecules, such as proteins and oligo-saccharides	[20, 22]
Microchip capillary electrophoresis	On-site soil inorganic nutrient content analysis	High sensitivity, simple extraction method	Limitations on the separation of ions at high concentrations	[25]
Agarose gel electrophoresis	Soil eDNA extraction	Easy isolation of separated DNA fragments from the gel, simple and rapid way to estimate quality, size, and length of separated DNA	More precise quality check and quantification require other technologies in combination	[66]

Table 1.
Summary of the applications, advantages, and limitations of different electrophoretic techniques in soil research.

promoting our understanding of different soil ecosystems. This knowledge can then be applied in ecosystem preservation and the search for beneficial interactions between different organisms in soil. The applications, advantages, and limitations of different electrophoretic techniques in soil research are summarized in **Table 1**.

Acknowledgements

This work was supported by grants from Science, Technology and Innovation Commission of Shenzhen Municipality (Shenzhen-Hong Kong-Macau Science and Technology Program, Category C, SGDX20210823103535007) and the Hong Kong Research Grants Council: Area of Excellence Scheme (AoE/M-403/16). J.Y. Chu copy-edited this manuscript. Any opinions, findings, conclusions, or recommendations expressed in this publication do not reflect the views of the Government of the Hong Kong Special Administrative Region or the Innovation and Technology Commission.

Author details

Cheuk-Hin Law[1], Long-Yiu Chan[1], Tsz-Yan Chan[1], Yee-Shan Ku[1,2]*
and Hon-Ming Lam[1,2,3]*

1 School of Life Sciences and Centre for Soybean Research of the State Key Laboratory of Agrobiotechnology, The Chinese University of Hong Kong, Hong Kong, China

2 Shenzhen Research Institute, The Chinese University of Hong Kong, Shenzhen, China

3 Institute of Environment, Energy and Sustainability, The Chinese University of Hong Kong, Hong Kong, China

*Address all correspondence to: ysamyku@cuhk.edu.hk; honming@cuhk.edu.hk

References

[1] Chen G, Zhu H, Zhang Y. Soil microbial activities and carbon and nitrogen fixation. Research in Microbiology. 2003;**154**:393-398

[2] He Z, Waldrip H, Wang Y. Applications of capillary electrophoresis in agricultural and soil chemistry research. In: He Z, editor. Capillary Electrophoresis: Fundamentals, Techniques and Applications. Hauppauge, NY: Nova Science Publishers; 2012

[3] Ewing AG, Wallingford RA, Olefirowicz TM. Capillary electrophoresis. Analytical Chemistry. 1989;**61**:292-303

[4] Virtanen R. Zone Electrophoresis in a Narrow-Bore Tube Employing Potentiometric Detection: A Theoretical and Experimental Study, Acta Polytechnica Scandinavica. Helsinki: Finnish Academy of Technical Sciences; 1974. ISBN 9789516660496

[5] McCord BR, Buel E. Capillary Electrophoresis in Forensic Genetics. 2nd ed. Amsterdam: Elsevier Ltd.; 2013. ISBN 9780123821652

[6] Kartsova LA, Bessonova EA. Biomedical applications of capillary electrophoresis. Russian Chemical Reviews. 2015;**84**:860-874

[7] Tatzber M, Klepsch S, Soja G, Reichenauer T, Spiegel H, Gerzabek MH. Determination of soil organic matter features of extractable fractions using capillary electrophoresis: An organic matter stabilization study in a carbon-14-labeled long-term field experiment. In: Labile Organic Matter—Chemical Compositions, Function, and Significance in Soil and the Environment. Madison, Wisconsin: Soil Science Society of America, Inc.; 2015. pp. 23-40

[8] Whatley H. Basic principles and modes of capillary electrophoresis. In clinical and forensic applications of capillary electrophoresis. In: Petersen JR, Mohammad AA, editors. Pathology and Laboratory Medicine. Totowa, NJ: Humana Press; 2003. pp. 21-58

[9] Gordon MJ, Huang X, Pentoney SL, Zare RN. Capillary electrophoresis. Science. 1988;**242**:224-228

[10] Olefirowicz TM, Ewing AG. Detection methods in capillary electrophoresis. In: Grossman PD, Colburn JC, editors. Capillary Electrophoresis. San Diego: Academic Press; 1992. pp. 45-85, ISBN 978-0-12-304250-7

[11] Schmitt-Kopplin P, Junkers J. Capillary zone electrophoresis of natural organic matter. Journal of Chromatography. A. 2003;**998**:1-20

[12] Vancampenhout K, Wouters K, Caus A, Buurman P, Swennen R, Deckers J. Fingerprinting of soil organic matter as a proxy for assessing climate and vegetation changes in last interglacial palaeosols (Veldwezelt, Belgium). Quaternary Research. 2008;**69**:145-162

[13] Pompe S, Heise K-H, Nitsche H. Capillary electrophoresis for a "finger-print" characterization of fulvic and humic acids. Journal of Chromatography. A. 1996;**723**:215-218

[14] Kinsey JL. Laser-induced fluorescence. Annual Review of Physical Chemistry. 1977;**28**:349-372

[15] Warren CR. Rapid and sensitive quantification of amino acids in soil extracts by capillary electrophoresis with laser-induced fluorescence. Soil Biology and Biochemistry. 2008;**40**:916-923

[16] Cheng C-N, Shufeldt RC, Stevenson FJ. Amino acid analysis of soils and sediments: Extraction and desalting. Soil Biology and Biochemistry. 1975;7:143-151

[17] Hamdan M, Righetti PG. Electrophoresis|capillary electrophoresis–mass spectrometry. In: Wilson ID, editor. Encyclopedia of Separation Science. Oxford: Academic Press; 2000. pp. 1188-1194. ISBN 978-0-12-226770-3

[18] Schmitt P, Garrison AW, Freitag D, Kettrup A. Capillary isoelectric focusing (CIEF) for the characterization of humic substances. Water Research. 1997;**31**:2037-2049

[19] Kovács P, Posta J. Separation of humic acids using capillary isoelectric focusing. Microchemical Journal. 2005;**79**:49-54

[20] Hancu G, Simon B, Rusu A, Mircia E, Gyéresi Á. Principles of micellar electrokinetic capillary chromatography applied in pharmaceutical analysis. Advanced Pharmaceutical Bulletin. 2013;**3**:1-8

[21] Zhang S, Yin X, Yang Q, Wang C, Wang Z. Determination of some sulfonylurea herbicides in soil by a novel liquid-phase microextraction combined with sweeping micellar electrokinetic chromatography. Analytical and Bioanalytical Chemistry. 2011;**401**:1071-1081

[22] Otsuka K, Terabe S. Micellar electrokinetic chromatography. Molecular Biotecnology. 1998;**9**:253-271

[23] Terabe S. Micellar electrokinetic chromatography. Analytical Chemistry. 2004;**76**:241A-246A

[24] Bindraban PS, Stoorvogel JJ, Jansen DM, Vlaming J, Groot JJR. Land quality indicators for sustainable land management: Proposed method for yield gap and soil nutrient balance. Agriculture, Ecosystems and Environment. 2000;**81**:103-112

[25] Smolka M, Puchberger-Enengl D, Bipoun M, Klasa A, Kiczkajlo M, Śmiechowski W, et al. A mobile lab-on-a-chip device for on-site soil nutrient analysis. Precision Agriculture. 2017;**18**:152-168

[26] Bardgett RD, van der Putten WH. Belowground biodiversity and ecosystem functioning. Nature. 2014;**515**:505-511

[27] Stein LY, Klotz MG. The nitrogen cycle. Current Biology. 2016;**26**:R94-R98

[28] Wardle DA. The influence of biotic interactions on soil biodiversity. Ecology Letters. 2006;**9**:870-886

[29] de Boer W, Folman LB, Summerbell RC, Boddy L. Living in a fungal world: Impact of fungi on soil bacterial niche development. FEMS Microbiology Reviews. 2005;**29**:795-811

[30] Fischer SG, Lerman LS. Length-independent separation of DNA restriction fragments in two-dimensional gel electrophoresis. Cell. 1979;**16**:191-200

[31] Børresen A-L, Hovig E, Brøgger A. Detection of base mutations in genomic DNA using denaturing gradient gel electrophoresis (DGGE) followed by transfer and hybridization with gene-specific probes. Mutation Research - Fundamental and Molecular Mechanisms of Mutagenesis. 1988;**202**:77-83

[32] Muyzer G, Teske A, Wirsen CO, Jannasch HW. Phylogenetic relationships of Thiomicrospira species and their identification in deep-sea hydrothermal vent samples by denaturing gradient gel electrophoresis of 16S rDNA

fragments. Archives of Microbiology. 1995;**164**:165-172

[33] Nakatsu CH. Soil microbial community analysis using denaturing gradient gel electrophoresis. Soil Science Society of America Journal. 2007;**71**:562-571

[34] Mahyarudin, Rusmana I, Lestari Y. Metagenomic of actinomycetes based on 16S rRNA and nifH genes in soil and roots of four Indonesian rice cultivars using PCR-DGGE. HAYATI Journal of Biosciences. 2015;**22**:113-121

[35] Hovig E, Smith-Sørensen B, Brøgger A, Børresen A-L. Constant denaturant gel electrophoresis, a modification of denaturing gradient gel electrophoresis, in mutation detection. Mutation Research Letters. 1991;**262**:63-71

[36] Mo Y. Probing the nature of hydrogen bonds in DNA base pairs. Journal of Molecular Modeling. 2006;**12**:665-672

[37] Muyzer G, de Waal EC, Uitterlinden AG. Profiling of complex microbial populations by denaturing gradient gel electrophoresis analysis of polymerase chain reaction-amplified genes coding for 16S rRNA. Applied and Environmental Microbiology. 1993;**59**:695-700

[38] Tsukuda M, Kitahara K, Miyazaki K. Comparative RNA function analysis reveals high functional similarity between distantly related bacterial 16S rRNAs. Scientific Reports. 2017;7:9993

[39] Nakatsu CH, Torsvik V, Øvreås L. Soil community analysis using DGGE of 16S rDNA polymerase chain reaction products. Soil Science Society of America Journal. 2000;**64**:1382-1388

[40] Norton JM, Alzerreca JJ, Suwa Y, Klotz MG. Diversity of ammonia monooxygenase operon in autotrophic ammonia-oxidizing bacteria. Archives of Microbiology. 2002;**177**:139-149

[41] Fujii T, Morimoto S, Hoshino Y, Okada H, Wang Y, Chu H, et al. Analysis of diversity and functions of microorganisms in soil using nucleic acids extracted from soil in NIAES. NIAES International Symposium Challenges for Agro-Environmental Research in Monsoon Asia. Review. 2009

[42] Jones CM, Thies JE. Soil microbial community analysis using two-dimensional polyacrylamide gel electrophoresis of the bacterial ribosomal internal transcribed spacer regions. Journal of Microbiological Methods. 2007;**69**:256-267

[43] Zijnge V, Welling GW, Degener JE, van Winkelhoff AJ, Abbas F, Harmsen HJM. Denaturing gradient gel electrophoresis as a diagnostic tool in periodontal microbiology. Journal of Clinical Microbiology. 2006;**44**:3628-3633

[44] Staley JT, Konopka A. Measurement of in situ activities of nonphotosynthetic microorganisms in aquatic and terrestrial habitats. Annual Review of Microbiology. 1985;**39**:321-346

[45] Ovreås L, Forney L, Daae FL, Torsvik V. Distribution of bacterioplankton in meromictic lake Saelenvannet, as determined by denaturing gradient gel electrophoresis of PCR-amplified gene fragments coding for 16S rRNA. Applied and Environmental Microbiology. 1997;**63**:3367-3373

[46] Gelsomino A, Keijzer-Wolters AC, Cacco G, van Elsas JD. Assessment of bacterial community structure in soil by polymerase chain reaction and denaturing gradient gel electrophoresis.

Journal of Microbiological Methods. 1999;**38**:1-15

[47] O'Farrell PH. High resolution two dimensional electrophoresis of proteins. The Journal of Biological Chemistry. 1975;**250**:4007-4021

[48] Klose J. Protein mapping by combined isoelectric focusing and electrophoresis of mouse tissues. Humangenetik. 1975;**26**:231-243

[49] Paul D, Nair SK. Stress adaptations in a plant growth promoting rhizobacterium (PGPR) with increasing salinity in the coastal agricultural soils. Journal of Basic Microbiology. 2008;**48**:378-384

[50] Benndorf D, Balcke GU, Harms H, von Bergen M. Functional metaproteome analysis of protein extracts from contaminated soil and groundwater. The ISME Journal. 2007;**1**:224-234

[51] Lin W, Wu L, Lin S, Zhang A, Zhou M, Lin R, et al. Metaproteomic analysis of ratoon sugarcane rhizospheric soil. BMC Microbiology. 2013;**13**:135

[52] Wu Y, Li CH, Zhao J, Xiao YL, Cao H. Metaproteome of the microbial community in paddy soil after long-term treatment with mineral and organic fertilizers. Israel Journal of Ecology & Evolution. 2015;**61**:146-156

[53] Wu L, Wang H, Zhang Z, Lin R, Zhang Z, Lin W. Comparative metaproteomic analysis on consecutively Rehmannia glutinosa-monocultured rhizosphere soil. PLoS One. 2011;**6**:e20611

[54] Magdeldin S, Enany S, Yoshida Y, Xu B, Zhang Y, Zureena Z, et al. Basics and recent advances of two dimensional-polyacrylamide gel electrophoresis. Clinical Proteomics. 2014;**11**:16

[55] Laemmli UK. Cleavage of structural proteins during the assembly of the head of bacteriophage T4. Nature. 1970;**227**:680-685

[56] Pawlowski J, Apotheloz-Perret-Gentil L, Altermatt F. Environmental DNA: What's behind the term? Clarifying the terminology and recommendations for its future use in biomonitoring. Molecular Ecology. 2020;**29**:4258-4264

[57] Kestel JH, Field DL, Bateman PW, White NE, Allentoft ME, Hopkins AJM, et al. Applications of environmental DNA (eDNA) in agricultural systems: Current uses, limitations and future prospects. Science of the Total Environment. 2022;**847**:157556

[58] van der Heyde M, Bunce M, Nevill P. Key factors to consider in the use of environmental DNA metabarcoding to monitor terrestrial ecological restoration. Science of the Total Environment. 2022;**848**:157617

[59] Andersen K, Bird KL, Rasmussen M, Haile J, Breuning-Madsen H, Kjaer KH, et al. Meta-barcoding of "dirt" DNA from soil reflects vertebrate biodiversity. Molecular Ecology. 2012;**21**:1966-1979

[60] Bienert F, De Danieli S, Miquel C, Coissac E, Poillot C, Brun JJ, et al. Tracking earthworm communities from soil DNA. Molecular Ecology. 2012;**21**:2017-2030

[61] Yoccoz NG, Brathen KA, Gielly L, Haile J, Edwards ME, Goslar T, et al. DNA from soil mirrors plant taxonomic and growth form diversity. Molecular Ecology. 2012;**21**:3647-3655

[62] Ogram A, Sayler GS, Barkay T. The extraction and purification of microbial DNA from sediments. Journal of Microbiological Methods. 1987;**7**:57-66

[63] Paul JH, Myers B. Fluorometric-determination of DNA in aquatic microorganisms by use of Hoechst-33258. Applied and Environmental Microbiology. 1982;**43**:1393-1399

[64] Somerville CC, Knight IT, Straube WL, Colwell RR. Simple, rapid method for direct isolation of nucleic-acids from aquatic environments. Applied and Environmental Microbiology. 1989;**55**:548-554

[65] Nagler M, Podmirseg SM, Ascher-Jenull J, Sint D, Traugott M. Why eDNA fractions need consideration in biomonitoring. Molecular Ecology Resources. 2022;**22**:2458-2470

[66] Lee PY, Costumbrado J, Hsu CY, Kim YH. Agarose gel electrophoresis for the separation of DNA fragments. Journal of Visualized Experiments. 2012;**20**:3923

[67] Voytas D. Agarose gel electrophoresis. Current Protocols in Molecular Biology. 2000;**51**:2.5A.1-23.3.6

[68] Zhou JZ, Bruns MA, Tiedje JM. DNA recovery from soils of diverse composition. Applied and Environmental Microbiology. 1996;**62**:316-322

[69] Roh C, Villatte F, Kim BG, Schmid RD. "In-gel patch electrophoresis": A new method for environmental DNA purification. Electrophoresis. 2005;**26**:3055-3061

[70] Roh C, Villatte F, Kim BG, Schmid RD. Comparative study of methods for extraction and purification of environmental DNA from soil and sludge samples. Applied Biochemistry and Biotechnology. 2006;**134**:97-112

[71] Brady SF. Construction of soil environmental DNA cosmid libraries and screening for clones that produce biologically active small molecules. Nature Protocols. 2007;**2**:1297-1305

[72] Li Y, Zhao S, Ma J, Li D, Yan L, Li J, et al. Molecular footprints of domestication and improvement in soybean revealed by whole genome re-sequencing. BMC Genomics. 2013;**14**:579

[73] Kim JH, Feng ZY, Bauer JD, Kallifidas D, Calle PY, Brady SF. Cloning large natural product gene clusters from the environment: Piecing environmental DNA gene clusters back together with TAR. Biopolymers. 2010;**93**:833-844

[74] Frostegård A, Courtois S, Ramisse V, Clerc S, Bernillon D, Le Gall F, et al. Quantification of bias related to the extraction of DNA directly from soils. Applied and Environmental Microbiology. 1999;**65**:5409-5420

[75] Aoshima H, Kimura A, Shibutani A, Okada C, Matsumiya Y, Kubo M. Evaluation of soil bacterial biomass using environmental DNA extracted by slow-stirring method. Applied Microbiology and Biotechnology. 2006;**71**:875-880

[76] Parachin NS, Schelin J, Norling B, Rådström P, Gorwa-Grauslund MF. Flotation as a tool for indirect DNA extraction from soil. Applied Microbiology and Biotechnology. 2010;**87**:1927-1933

[77] Cristescu ME, Hebert PDN. Uses and misuses of environmental DNA in biodiversity science and conservation. Annual Review of Ecology, Evolution, and Systematics. 2018;**49**(49):209-230

[78] Guerrieri A, Carteron A, Bonin A, Marta S, Ambrosini R, Caccianiga M, et al. Metabarcoding data reveal vertical multitaxa variation in topsoil communities

during the colonization of deglaciated forelands. Molecular Ecology. 2022;**32**(23):6304-6319

[79] Rota N, Canedoli C, Ferre C, Ficetola GF, Guerrieri A, Padoa-Schioppa E. Evaluation of soil biodiversity in alpine habitats through eDNA metabarcoding and relationships with environmental features. Forests. 2020;**11**:738

[80] Taberlet P, Prud'homme SM, Campione E, Roy J, Miquel C, Shehzad W, et al. Soil sampling and isolation of extracellular DNA from large amount of starting material suitable for metabarcoding studies. Molecular Ecology. 2012;**21**:1816-1820

[81] Koshel BM, Wirth MJ. Trajectory of isoelectric focusing from gels to capillaries to immobilized gradients in capillaries. Proteomics. 2012;**12**:2918-2926